The World through the Lens of Mathematics

Hundreds of meticulously crafted mathematical problems and puzzles in this book are incorporated into fascinating stories about our world. These wor(l)d problems are grouped by their mathematical concepts such that the titles of sections and chapters combine both mathematical and applied terms, hinting at the topics covered. Detailed solutions conclude each section.

Following in the success of the authors' previous book, *USA Through the Lens of Mathematics*, this text contributes to the novel pedagogical call for a more multidisciplinary approach in education. The various types of questions posed in *The World through the Lens of Mathematics* are stimulating, entertaining, and educational. Their main objective is to provide a thorough review of the fundamental concepts of algebra and geometry, reduce mathematical anxiety, and emphasize the applicability and versatility of mathematics.

Working these problems shatters the barriers between the students and mathematics by encouraging them to look at the subject from a different perspective. Students are simultaneously enriched with new knowledge of historical events, customs, and geography of countries around the world, each one of which is mentioned throughout the pages of this book.

The World through the Lens of Mathematics

Natali Hritonenko and Yuri Yatsenko

CRC Press is an imprint of the
Taylor & Francis Group, an **informa** business

First edition published 2025
by CRC Press
2385 NW Executive Center Drive, Suite 320, Boca Raton FL 33431

and by CRC Press
4 Park Square, Milton Park, Abingdon, Oxon, OX14 4RN

CRC Press is an imprint of Taylor & Francis Group, LLC

Library of Congress Cataloging-in-Publication Data
Names: Hritonenko, Natali, author. | Yatsenko, Y. P. (Yurii Petrovich), author.
Title: The world through the lens of mathematics / authored by
Natali Hritonenko and Yuri Yatsenko.
Description: First edition. | Boca Raton, FL : CRC Press, 2025. |
Includes bibliographical references and index.
Identifiers: LCCN 2024003012 | ISBN 9781032398617 (hbk) |
ISBN 9781032398594 (pbk) | ISBN 9781003351702 (ebk)
Subjects: LCSH: Mathematics–Miscellanea. | Civilization–Miscellanea. |
Interdisciplinary approach in education.
Classification: LCC QA99 .H837 2025 | DDC 510–dc23/eng/20240409
LC record available at https://lccn.loc.gov/2024003012

ISBN: 9781032398617 (hbk)
ISBN: 9781032398594 (pbk)
ISBN: 9781003351702 (ebk)

DOI: 10.1201/9781003351702

Typeset in Palatino
by Newgen Publishing UK

Contents

Preface

We greatly appreciate your interest in this book, the sequel to our successful previous work, *USA Through the Lens of Mathematics.* Both books were written for and inspired by educators and students, regardless of whether they are already math lovers or soon to be! This book shows the magic, versatility, and applicability of mathematics through carefully crafted word problems incorporated into fascinating stories from around the world.

When we first envisioned this book, we wanted to design mathematical problems that bring together history, heritage, geography, and traditions from countries around the globe. It is impossible to give a comprehensive coverage of the whole story behind the world in one book; thus, we have not attempted to do so. Instead, we have decided to summarize meaningful events and present related problems or puzzles to maintain a delicate balance of keeping this book engaging and meaningful but not too broad. Every single country of the world is mentioned throughout its pages.

Meticulously designed problems and puzzles of different types and complexity are grouped together in sections and chapters following their mathematical content, but all of them are accompanied by a relevant applied outline. We hope that you will find these carefully sharpened tasks stimulating, entertaining, and educational, while also fulfilling their main objective of providing a thorough review of basic mathematical foundations taught in middle to high school.

We have worked hard to ensure that this creative, interdisciplinary approach will promote a deeper understanding of mathematics, enhance logical skills, and motivate the simultaneous learning of mathematics, social studies, and deeply moving stories of the world. We invite you to travel the globe, enjoy its astonishing stories, and explore unique natural and manmade monuments while solving the wonderful problems and puzzles offered in this book.

You will not be disappointed, as this book is for curious people like you. We hope that you will enjoy working on these problems as much as we enjoyed making them!

If you have any questions or comments, please do not hesitate to contact us. Thank you.

Introduction

This entertaining book aims to show the beauty, magic, and power of mathematics, along with its applicability and versatility. The purpose of this book is to contradict the popular belief that mathematical subjects are "boring" and consist of a "set of unrelated statements irrelevant to our life outside of the class."

Considered among the most challenging subjects in high schools and colleges, mathematics courses have become a barrier for successful studies by many students around the globe. Their fear is caused by insufficient understanding of basic mathematical concepts, which, in turn, essentially form the alphabet of general sciences required to succeed in physics, chemistry, biology, engineering, business, and other subjects. This book lessens this struggle with mathematics by offering amusing non-traditional word problems that spark interest and inspire readers to solve them. Integrating such thoroughly designed problems into multi-disciplinary, cross-cultural content enhances curiosity, boosts the development of strong mathematical and logical skills, and broadens understanding of the world. This book follows the modern educational trend to promote interdisciplinary research. Algebraic and geometric fundamentals are reviewed while presenting hundreds of fascinating facts about the world.

This book has been inspired by our students and the educators we work with and participants of our numerous special sessions and workshops on innovative educational techniques. We have received many requests to extend our ideas presented in our previous book *USA Through the Lens of Mathematics.*

Uniqueness of the Book

This fun book enriches modern, innovative educational strategies for the simultaneous learning of mathematics and social studies. Its *uniqueness* is in the unexpected combination of mathematics and deeply moving stories of the world incorporated into various problems and puzzles. The *multidisciplinary* nature of this book is another highlight that distinguishes it from other educational mathematics books on the market. Its captivating problems and puzzles explore the history and geography of the world, the heritage and development of its countries, and both excavations of the past and new discoveries. Elements of surprise will trigger a positive emotional response, which, in turn, will make readers intrigued, curious, and inspired to improve their mathematical foundations.

This book is not a textbook but, rather, an excellent supplementary text for universities, colleges, and schools that can be used in different subjects, from physics and chemistry to history and literature. Arranged following their mathematical content, the problems cover and review the basic algebraic and geometric formulas and rules widely used across many disciplines.

Audience

This book shows the versatility of mathematics in addressing many diverse real-life issues and emphasizes the greatness of interdisciplinary approaches in education. Its problems can be effectively incorporated into different subjects, used in a classroom, or completed during leisure time.

This book offers school, college, and university students a rich collection of fun and unique problems that will assist them in reviewing fundamental algebraic rules, enriching their mathematical literacy, enhancing their logical skills, and, simultaneously, advancing their understanding of the history and geography of the world and its countries. This will help them become better prepared for not only their exams, but also their future studies.

Working on the advanced and often complex problems and puzzles presented in this book, college seniors and pre-service teachers will not only enhance their mathematical literacy but also master their polymath and logical skills, thus equipping themselves with knowledge needed for successfully passing their teacher certification exams and beyond.

School curricula are shifting toward cross-cultural and cross-disciplinary teaching in a rapidly changing world. The problems of this book touch different aspects of our society and surroundings, and will undoubtedly provide a valuable source for educators for developing interdisciplinary curricula and motivating global awareness as well. Instructors that seek innovative ways to demonstrate the relevance of mathematics to other subjects can enliven their classes with ready-made entertaining problems that make mathematics much more enjoyable to their students, even fun! The detailed solutions and alternative methods included in this book are a great asset for teachers of all levels and disciplines.

The variety of difficulty levels of suggested problems also make them suitable for clubs and competitions across many student levels and subjects.

Finally, the authors hope that this book will be interesting for everyone who loves adventure, challenge, and enjoys mathematics.

Structure of the Book

This interdisciplinary book consists of 24 sections, arranged in five chapters and an index. The titles of the chapters and sections combine mathematical and applied terms, hinting at the topics provided. Following the valuable comments that we received from educators and students about our previous book, *USA Through the Lens of Mathematics*, the problems are grouped based on their mathematical content rather than on their applied topic:

- The first chapter reviews fundamental rules of algebra.
- The second chapter concentrates on creating and solving linear and nonlinear equations and their systems.
- The third chapter offers a variety of word problems.
- The fourth chapter reviews basic formulas of geometry.
- The fifth chapter contains various logical puzzles.

Each section consists of ten problems and aims to reveal both amazing facts about our world and enhance the understanding of basic mathematical properties.

The *answer key* follows each problem, and *detailed solutions* conclude each section. Alternate methods of solving problems are presented if each way involves different techniques or nonstandard methods. A brief reference list concludes each section.

Through researching and collecting information for this book, the authors studied various sources, reference works, and websites. In some cases, these sources may have reported slightly different data; despite these inconsistencies, this book provides a reliable general description of real stories and events.

The *Index* at the end of the book lists the countries mentioned in the book in relation to their respective problems.

Mathematical Problems

This book has been curated to fit any mathematical abilities, backgrounds, and tastes. A brief real-life fact or event accompanies each problem, though it does not provide information relevant to its solution. Each problem itself has all the information and data needed for its solution and can be considered independently; however, each mathematical problem complements its related story. The numbering of formulas is avoided to make the explanations easier to understand.

The problems are of different styles, types, and complexity. Several pairs of problems may look similar, though a small variation in their formulation changes the level of their difficulty and requires a completely different solution strategy. Moreover, some problems cannot be solved based on the information provided or their solution lacks any applied interpretation.

There are three design styles of problems throughout this book:

- *Word problems* vary in their presentation. Some of them provide an equation to be solved or a mathematical statement to be simplified, others involve creating and solving mathematical equations, inequalities, or their system. Choosing proper unknowns and solving obtained equations can be challenging in some problems but straightforward in others. Most of these problems require a single answer, though there are some multiple-choice problems with several possible answers to choose from.
- *Problems with hints* contain additional information as extra clues or tips. A hint can:
    - lead to a statement needed to solve a problem,
    - guide to appropriate information,
    - be reduced to a statement already presented in the problem,
    - help to find a unique solution,
    - be useless or not applicable to the problem.

It is an art to select the right hints and explain why other hints are useless or lead to either no or many solutions.

- *Logical puzzles*, like sudoku, kakuro, inky, magic square, logic, triangular, pattern, de-coding, and others, are offered to be solved to reveal some amazing information. All of them involve a simple manipulation of mathematical properties.

Independent of their design, the problems are of varying levels of complexity not shown to avoid unnecessary fear of inability of solving, or even starting, problems marked as challenging.

There are three levels of complexity:

- *Easy problems* can be figured out using straightforward application of mathematical relationships, or even without employing any mathematical formulas.
- *Moderately difficult problems* involve manipulation and combinations of mathematical formulas, several solution steps, and logical thinking if not all necessary information is provided, or choosing an appropriate answer if several solutions are obtained.

- *Challenging problems* require deeper thinking, critical reasoning, and thorough investigation to reveal surprising mathematical connections that lead to the solution. A certain level of mathematical background is desired to solve the problems of this group.

To illustrate the variety of mathematical techniques, different ways of solving a problem are often shown and thoroughly analyzed. The reasoning of hint selection is justified and discussed for problems with hints.

The authors hope that this variety of problems can be effectively used to enhance mathematical knowledge and skills, reduce mathematics anxiety, and enhance a deeper interdisciplinary expertise through the practice of learning new techniques and strategies.

About the Authors

Dr. Natali Hritonenko is an award-winning professor of mathematics and associate dean at Prairie View A&M University. During her prolific career, Dr. Hritonenko has authored 7 books and over 150 papers. Her books are used as textbooks and have been translated. Dr. Hritonenko is continuously working on improving mathematical education and is the founder and director of PVAMU REU Site: Mathematical Modeling in the Sciences, supported by the NSF and Mellon Initiative for Inclusive Faculty Excellence.

Dr. Yuri Yatsenko holds a PhD from Kiev National University, Ukraine and DSc from the USSR Academy of Sciences, Moscow. He has been a professor at science, business, and technology schools in Ukraine, Poland, Kazakhstan, Belgium, Canada, and the US. Since 2002, he has been a professor at Houston Baptist University. Dr. Yatsenko has published 8 books and more than 200 papers in top-ranked scientific journals.

Acknowledgments

The authors would like to express gratitude to their students, teachers, and colleagues for friendly critique, useful suggestions, inspirations, and ideas. A special thanks goes to the former students, collaborators, and educators the authors have been working with, to seventh grade student Mr. Philip Alexander and fifth grade student Mr. Andrew Neel for testing all puzzles and commenting on them, and to their families for their incredible support. Their endorsement and encouragement are tremendous and cannot be expressed in any words.

While researching milestones and pivotal moments presented in the book, the authors used a wide array of wonderful electronic reference tools including Wikipedia and other websites. The authors are sincerely grateful to professionals for creating and maintaining these web sources and making them available to the general audience.

1

Our Algebraic World

> Mathematics is a game played according to certain simple rules with meaningless marks on paper.
>
> ~ *David Hilbert* (1862–1943)[1]

What point on Earth is the closest to the stars? What are the largest, longest, and deepest rivers on the planet? How much have borders between countries changed over time? What unions of countries do you know? What are UNESCO sites and where are they? Who created the valuable inventions still widely used today? When and where did the Olympic Games start?

You will find the answers to these and other intriguing questions and discover fascinating facts about our world while solving the 60 problems of 6 sections of this chapter. At the same time, you will review the basic concepts of fractions and factorization, enhance your knowledge of sequences and optimization, and master the main relations of logarithm and probability. This well-thought blend of mathematical and real-world concepts aims to solidify the algebraic fundamentals widely used in all disciplines. Lose the idea that algebra is just a boring set of rules and methods to remember and, instead, experience the inspirational gateway that algebra provides to explore the world around you. *Bon voyage!*

1.1 Our World in Fractions

Divided Countries

Empires and countries appeared and disappeared, were united and divided, changing their borders and forming new states. New names of countries arose on the world map. North, South, East, West are often added to the former country, like East Germany and West Germany, South Somalia and Somalia, South Korea and North Korea.

[1] Hilbert spaces, numbers of the form $4n + 1$, the Hilbert cube, and many other important theorems, problems, and paradoxes are named after this famous German mathematician.

DOI: 10.1201/9781003351702-1

For centuries before its division, the Korean Peninsula, bordered with Russia and China and surrounded by the Yellow Sea, Sea of Japan, Korea Bay, and Korea Strait, was a single unified Korea ruled by generations of dynasties. It was occupied by Japan for over 35 years after the Russo-Japanese War in 1905. The division of the Korean Peninsula into North and South Korea was due to unexpected casualties of the Cold War between two rival superpowers, the Soviet Union and the United States.

Problem. The Korean Peninsula was divided to North and South Korea in the 20th century. The last two digits of the year in the way they are appeared in the year are the denominator of an irreducible fraction representation of the infinite repeating decimal 0.8444444… . The numerator of the fraction stands for the parallel of the border between North Korea and South Korea. Find the year when the Korean Peninsula was divided into North and South Korea and the location of the border.

Answer. 1945, 38

Reunited Countries

Countries and empires are established, split, and re-united. East Germany and West Germany, South Vietnam and North Vietnam are just a few examples of countries that have come together after years of separation.

The Socialist Republic of Vietnam is at the eastern edge of Southeast Asia bordering with China, Laos, and Cambodia. The first people settled on the Red River Delta several millenniums ago. The Chinese Han dynasty conquered the land from 111 BC until 939 when the first Vietnamese dynasty emerged absorbing Chinese influence and religion. The last imperial dynasty fell to the French in 1887. Independence from France was proclaimed by revolutionary Ho Chi Minh in 1945. It was decided at the Geneva conference that Vietnam would be temporarily divided along the Ben Hai River separating the French and the Viet Minh military forces until the democratic election.

Problem. Vietnam was split into North and South Vietnam in the 20th century and re-united as one country 22 years later. The last two digits of the year (as they appear in the year) the country was divided are the denominator of a irreducible fraction representation of the infinite repeating decimal 0.3148148148… . The numerator of the fraction stands for the parallel of the set-up border between North Vietnam and South Vietnam. Find the years when Vietnam was divided into North and South Vietnam and then re-united, and the location of the border. Is 0.3148148148… . irrational?

Answer. 1954, 1976, 17, no

Country Capitals

The capital of a country, state, kingdom, or region is its most important city, though it is not necessarily its largest city. Most countries have one capital, while some countries have several capitals used for different purposes and caused by different causes. For instance, when Paris became the capital for the German military administration during World War II, Vichy served as de facto administrative capital of France. Quezon City in Philippines was its official capital in 1948–1976, but Manila was its de facto seat of government at that time. Do you know any other countries with multiple capitals?

Problem. Find the number of countries that have more than one capital presented as the whole number of the improper fraction

$$\frac{1}{\dfrac{1}{2+\dfrac{1}{\frac{2}{5}-\frac{1}{3}}}+\dfrac{2}{3+\dfrac{7}{\frac{5}{6}-\frac{1}{3}}}}$$

presented as a mixed fraction, and the smallest and largest numbers of capitals that these countries currently have presented as the numerator and denominator of this mixed fraction.

Answer. 5, 2, 3

New Countries

Political, cultural, and religious distinctions between different ethnic groups of the citizens of Yugoslavia and the Soviet Union are among the variety of reasons that led to the breakup of these two powerful countries at the end of the 20th century.

Yugoslavia, meaning the *South Slavic Land*, was a country in the Balkan Peninsula in Southeast and Central Europe. Some territories of the former Austro-Hungarian Empire unified with the Kingdom of Serbia to form the Kingdom of the Serbs, Croats, and Slovenes under King Peter I of Serbia. The kingdom gained international recognition in 1922 at the Conference of Ambassadors in Paris. The Kingdom of Yugoslavia was renamed as the Socialist Federal Republic of Yugoslavia after World War II but later collapsed, creating new countries.

The Soviet Union, or the Union of Soviet Socialist Republics (USSR), located in both Europe and Asia, traces its origin to the 1917 October Revolution when a far left-political party led by Vladimir Ilyich Lenin took the power in Russia and established the world's first constitutionally guaranteed socialist state, the RSFSR. The Declaration of the Creation of the USSR formed by

Russia, Ukraine, Belarus, and the Transcaucasia was signed in 1922. The Transcaucasia was divided into Georgia, Armenia, and Azerbaijan in 1936. The United States recognized the Soviet Union in 1936, and the country was admitted to the League of Nations in 1934. New independent countries were formed after the dissolution of the USSR in December 1991.

Problem. Find the numbers of countries that came from Yugoslavia and the Soviet Union if these numbers are the numerator and denominator of the expression

$$\frac{1}{2+\frac{6}{5}+\frac{18}{25}+\frac{54}{125}+\frac{162}{625}+\frac{486}{3125}+\ldots}+\frac{1}{4-\frac{4}{3}+\frac{4}{9}-\frac{4}{27}+\frac{4}{81}-\frac{4}{243}+\ldots}$$
$$-\frac{1}{(2+\frac{6}{5}+\frac{18}{25}+\frac{54}{125}+\frac{162}{625}+\frac{486}{3125}+\ldots)\cdot(4-\frac{4}{3}+\frac{4}{9}-\frac{4}{27}+\frac{4}{81}-\frac{4}{243}+\ldots)},$$

written as one unreducible fraction.

Answer. 7, 15

National Anthem

The national anthem of a country is one of its most significant symbols. As the national flag, the anthem serves as an expression of the country's identity. Anthems are performed at the public appearances of a country's head or monarch, during the Olympics, and other important occasions. The national anthem is commonly written in a de facto or official language of the country and may have several versions in countries with multiple languages, though the national anthem of some countries is silent with only musical tunes.

Problem. Each verse of one of two official national anthems of New Zealand is performed in different languages. This number of languages is one less than the value of the numerator of the sum of the infinite series

$$\frac{1}{3}+\frac{1}{8}+\frac{1}{15}+\frac{1}{24}+\frac{1}{35}+\frac{1}{48}+\ldots.$$

The denominator of the sum reveals the number of countries that do not have any official lyrics in the anthem. The number of languages used to perform South Africa's national anthem is one more than the value of the denominator. Find the number of countries with these unique features of their national anthems. List them.

Answer. 2, 4, 5

1.2 National Flags

A national flag is one of the most important symbols of a state recognizable in the world. Its design and colors signify the history, culture, religion, and heritage of the nation. Most dependencies and former colonies use the same proportions of their flags as their mother countries. For instance, the bordering ratio of all British Overseas Territories is 1:2 ratio.

Nepal is the only nation with a national flag in the shape of two stacked triangles or a double pennon. All other national flags are rectangular with various proportions between its width and length: the proportion of 15 to 22 is used by the Bolivian national flag, 8 to 11 is used by the Israeli flag, and 10 to 19 is used by Marshall Islands. The most common ratio of host to fly is 2:3 that used by 85 of 195 sovereign states, followed by 1:2 used by 54 countries, 3:5, and so on. Togo is the only country that followed the golden ratio while designing its national flag.

Problem. Express the fly (length) l of a flag and its host (width) w in terms of the flag area A if a bordering ratio of a rectangular national flag is a to b.

Express the fly and host of the national flag of Togo via the area A, if a bordering ratio of a rectangular national flag proportion is $1:\frac{1+\sqrt{5}}{2}$.

Answer. $\sqrt{\frac{Aa}{b}}, \sqrt{\frac{Ab}{a}}, \sqrt{\frac{A\cdot(\sqrt{5}-1)}{2}}, \sqrt{\frac{A\cdot(\sqrt{5}+1)}{2}}$

Skyscrapers

Egyptian pyramids, European cathedrals, Muslim minarets, the Lighthouse of Alexandria, and other constructions that go up in the sky have always stimulated and encouraged humanity to go higher. Buildings with ten floors were so unique in the 1880s that they were called *skyscrapers*. Technological progress goes extremely fast supporting architectural design that was impossible to imagine a century ago and changing the definition of skyscrapers to habitable high-rise buildings with over 40 floors and over 150 meters or 492 feet tall. China, the United States, and the United Arab Emirates have the most skyscrapers. With over 3000 skyscrapers, Shanghai, Shenzhen, and Hong Kong are way above any other country in the world. Mumbai in India, Tokyo in Japan, and Kuala Lumpur in Malaysia are also among the top ten cities with the most skyscrapers.

Problem. What rank in the list of the cities with the most skyscrapers does Hong Kong take if it is defined as

$$\frac{\sqrt[3]{2+\sqrt{5}}+\sqrt[3]{2-\sqrt{5}}}{\sqrt{2}\cdot\sqrt{3-\sqrt{5}}-\sqrt{5}}?$$

Evaluate without a calculator.

Answer. 1

Shared Islands

Islands are commonly a territory of one nation but some of them, like Papua New Guinea or Timor, are governed by several countries. The international border between Finland and Sweden goes through Märket and Kataja Islands. Tierra del Fuego belongs to Chile and Argentina. The world's third largest island and the Asian largest island Borneo is the only island that is shared by three countries, Malaysia and Brunei to the north in the Northern Hemisphere, and Indonesia to the south in the Southern Hemisphere. Sovereign country-sultanate Brunei is entirely on Borneo.

Problem. Indonesia Malaysia and Brunei share the beautiful mountainous island of Borneo. The Malaysian territory of Borneo is 192 thousand square kilometers more than the one that belongs to Brunei but 342 thousand square kilometers less than the Indonesian part that that takes 72.58% of the island. Find the total area of Borneo in thousand km^2, and its areas in thousand km^2 and % that belong to Indonesia, Malaysia, and Brunei.

Answer. 744 thousand km^2, 540 thousand km^2, 72.58%,
198 thousand km^2, 26.62%, 6 thousand km^2, 0.8%

Feeding Animals

Cut and dried grass, legumes, flowers, or other foliage, called hay, are stored to feed large grazing animals and smaller domesticated animals during winter, drought, and when pasture is unavailable.

Problem. A Kansas farm in the US has enough hay to feed the cows for x days, the goats for y days, and the sheep for z days. How much time does this hay last to feed all animals?

A. xyz

B. $x+y+z$

C. $\frac{1}{x}+\frac{1}{y}+\frac{1}{z}$

D. $\frac{1}{x+y+z}$

E. $\frac{1}{xyz}$

F. $\frac{x+y+z}{xyz}$

G. $\frac{xy+xz+yz}{xyz}$

H. $\frac{xyz}{xy+xz+yz}$

Answer. H

Healthy Weight

Obesity is a modern epidemic in most parts of the world. It increases the risk of heart disease, stroke, and high cholesterol, and causes other health problems. Federal agencies have issued guidelines on defining a healthy weight range. The Body Mass Index (BMI), the Body Surface Area (BSA), and the Waist-to-Height Ratio (WHtR) of a person are commonly used to identify the weight category: underweight, healthy, overweight, or obese. These measures are good indicators for an average person, though their limitations lead to various formulas for their calculation. The ratio of a person's weight in kilograms to their height in meters squared, the BMI is adopted by most countries and international agencies.

The growth of unhealthy fast and processed food, different lifestyles, genetics, dietary, environmental, and cultural factors are among many reasons that cause obesity. Indeed, Japan with a diet concentrated on seafood, freshness, and modest portions is among the countries with the least obesity and the highest life expectancy.

Following BMI-based international reports, Madagascar, Eritrea, and Ethiopia are the least obese countries, while the Cook Islands, Nauru, and Palau lead the list of the most obese countries having over 50% of adults in this category. Actually, countries in the South Pacific have the highest obesity rate.

Problem. Find the upper value of BMI of a person in the healthy range if it is 100 times the proper fraction that doubles if 2 is added to its numerator and denominator.

Answer. 25%

Solutions

Divided Countries. The infinite repeating decimal or recurring decimal 0.8444444... consists of one 8 and 4 repeated infinitely many times. Multiplying the number 0.8444444… by 100 and then multiplying 0.8444444… by 10 we obtain two equalities

$$100\cdot 0.8444444\ldots = 84.44444\ldots,$$
$$10\cdot 0.8444444\ldots = 8.44444\ldots.$$

Subtracting the second equality from the first one we get $90\cdot 0.8444444\ldots = 76$, from which follows

$$0.844444\ldots = \frac{76}{90} = \frac{38}{45}.$$

The Korean Peninsula was divided into North and South Korea around the 38th parallel on August 10, 1945.

Reunited Countries. The infinite repeating decimal or recurring decimal 0.3148148148… consists of one-digit 3 and a three-digit group 148 repeated infinitely many times. Multiplying the number 0.3148148148… by 10,000 and then multiplying 0.3148148148… by 10, we obtain the following two equalities

$$10000\cdot 0.3148148148\ldots = 3148.148148\ldots,$$
$$10\cdot 0.3148148148\ldots = 3.148148148\ldots.$$

Subtracting the second equality from the first one we get $9990\cdot 0.3148148148\ldots = 3145$, from which follows that

$$0.3148148148\ldots = \frac{3145}{9990} = \frac{17}{54}.$$

North and South Vietnam were merged on July 2, 1976 to form the Socialist Republic of Viet Nam after being split on July 21, 1954, with the border around the 17th parallel.

The number 0.3148148148… . is a *ratio*nal number because it can be presented as the *ratio* 17 to 54.

Country Capitals. Let us simplify the fraction as

$$\frac{1}{\dfrac{1}{2+\dfrac{1}{\dfrac{2}{5}-\dfrac{1}{3}}}+\dfrac{2}{3+\dfrac{7}{\dfrac{5}{6}-\dfrac{1}{3}}}} = \frac{1}{\dfrac{1}{2+\dfrac{1}{\dfrac{6-5}{15}}}+\dfrac{2}{3+\dfrac{7}{\dfrac{5-2}{6}}}} = \frac{1}{\dfrac{1}{2+\dfrac{1}{\dfrac{1}{15}}}+\dfrac{2}{3+\dfrac{7}{\dfrac{3}{6}}}}$$

$$=\frac{1}{\frac{1}{2+15}+\frac{2}{3+14}}=\frac{1}{\frac{1+2}{17}}=\frac{17}{3}=5\frac{2}{3}.$$

Five countries, Bolivia (Sucre and La Paz), Eswatini (Mbabane and Lobamba), Malaysia (Putrajaya and Kuala Lumpur), Sri Lanka (Colombo and Sri Jayawardenepura Kotte), and South Africa (Pretoria, Cape Town, and Bloemfontein) currently have multiple capitals.

New Countries. Let us rewrite the first term of the expression as

$$\frac{1}{2+\frac{6}{5}+\frac{18}{25}+\frac{54}{125}+\frac{162}{625}+\frac{486}{3125}+\ldots}$$

$$=\frac{1}{2+2\cdot\left(\frac{3}{5}\right)^1+2\cdot\left(\frac{3}{5}\right)^2+2\cdot\left(\frac{3}{5}\right)^3+2\cdot\left(\frac{3}{5}\right)^4+2\cdot\left(\frac{3}{5}\right)^5+\ldots}.$$

It is easy to see that the denominator is an infinite decreasing geometric series with the first term $a = 2$ and the ratio $r=\frac{3}{5}<1$. Then the sum of all terms is $S=\frac{a}{1-r}=\frac{2}{1-\frac{3}{5}}=5$ and

$$\frac{1}{2+\frac{6}{5}+\frac{18}{25}+\frac{54}{125}+\frac{162}{625}+\frac{486}{3125}+\ldots}=\frac{1}{5}.$$

Performing similar steps with the second term with the denominator presented as an infinite decreasing alternating geometric series with the first term $a = 1$ and the ratio $r=-\frac{1}{3}$, we get

$$\frac{1}{4-\frac{4}{3}+\frac{4}{9}-\frac{4}{27}+\frac{4}{81}-\frac{4}{243}+\ldots}=\frac{1}{4\cdot\left(1-\frac{1}{3}+\frac{1}{3^2}-\frac{1}{3^3}+\frac{1}{3^4}-\frac{1}{3^5}+\ldots\right)}$$

$$=\frac{1}{4\cdot\frac{1}{1-\left(-\frac{1}{3}\right)}}=\frac{1}{3},$$

and then

$$\frac{1}{(2+\frac{6}{5}+\frac{18}{25}+\frac{54}{125}+\frac{162}{625}+\frac{486}{3125}+...)\cdot(4-\frac{4}{3}+\frac{4}{9}-\frac{4}{27}+\frac{4}{81}-\frac{4}{243}+...)}=\frac{1}{3\cdot 5}=\frac{1}{15}.$$

Thus,

$$\frac{1}{2+\frac{6}{5}+\frac{18}{25}+\frac{54}{125}+\frac{162}{625}+\frac{486}{3125}+...}+\frac{1}{4-\frac{4}{3}+\frac{4}{9}-\frac{4}{27}+\frac{4}{81}-\frac{4}{243}+...}$$

$$-\frac{1}{(2+\frac{6}{5}+\frac{18}{25}+\frac{54}{125}+\frac{162}{625}+\frac{486}{3125}+...)\cdot(4-\frac{4}{3}+\frac{4}{9}-\frac{4}{27}+\frac{4}{81}-\frac{4}{243}+...)}$$

$$=\frac{1}{5}+\frac{1}{3}-\frac{1}{15}=\frac{7}{15}.$$

Yugoslavia split into 7 countries, Bosnia and Herzegovina, Croatia, Kosovo, Macedonia, Montenegro, Serbia, and Slovenia.

15 countries, Armenia, Azerbaijan, Belarus, Estonia, Georgia, Kazakhstan, Kyrgyzstan, Latvia, Lithuania, Moldova, Russia, Tajikistan, Turkmenistan, Ukraine, and Uzbekistan, used to be republics of the Soviet Union.

National Anthem. Let us rewrite

$$\frac{1}{3}+\frac{1}{8}+\frac{1}{15}+\frac{1}{24}+\frac{1}{35}+\frac{1}{48}+....\text{as}$$

$$\frac{1}{4-1}+\frac{1}{9-1}+\frac{1}{16-1}+...+\frac{1}{n^2-1}+....=\frac{1}{2^2-1}+\frac{1}{3^2-1}+\frac{1}{4^2-1}+...+\frac{1}{n^2-1}+....$$

that using $\frac{1}{n^2-1}=\frac{1}{2}\left(\frac{1}{n-1}-\frac{1}{n+1}\right)$, can be reshaped as

$$\frac{1}{2}\left(\left(\frac{1}{1}-\frac{1}{3}\right)+\left(\frac{1}{2}-\frac{1}{4}\right)+\left(\frac{1}{3}-\frac{1}{5}\right)+\left(\frac{1}{4}-\frac{1}{6}\right)+..+\left(\frac{1}{n-1}-\frac{1}{n+1}\right)+...\right)=\frac{3}{4}.$$

One of New Zealand's two national anthems is sung with the first verse *Aotearoa* in Māori and the second verse *God Defend New Zealand* in English. The national anthems of Spain, Bosnia and Herzegovina, Kosovo, and San Marino do not have any official lyrics. South Africa's national anthem puts two different songs together and performs it in five of the country's 11 official languages.

National Flags. The ratio between the width w and the length l of the flag is $w/l = a/b$ and its area $A = wl = w^2b/a$. Thus,

$$w = \sqrt{\frac{Aa}{b}} \text{ and } l = \sqrt{\frac{Ab}{a}}.$$

For the flag of Togo:

$$w = \sqrt{\frac{A \cdot 2}{1+\sqrt{5}}} = \sqrt{\frac{A \cdot 2(1-\sqrt{5})}{(1+\sqrt{5})(1-\sqrt{5})}} = \sqrt{\frac{A \cdot (\sqrt{5}-1)}{2}} \text{ and } l = \sqrt{\frac{A \cdot (\sqrt{5}+1)}{2}}.$$

Skyscrapers. Let us denote $x = \sqrt[3]{2+\sqrt{5}} + \sqrt[3]{2-\sqrt{5}}$. Applying the formulas $(a + b)^3 = a^3 + 3a^2b + 3ab^2 + b^3$ and $(a - b)(a + b) = a^2 - b^2$, we can cube the expression and simplify it as

$$x^3 = (\sqrt[3]{2+\sqrt{5}} + \sqrt[3]{2-\sqrt{5}})^3 = 2+\sqrt{5} + 3 \cdot (\sqrt[3]{2+\sqrt{5}})^2 \cdot \sqrt[3]{2-\sqrt{5}} + 3 \cdot \sqrt[3]{2+\sqrt{5}}(\sqrt[3]{2-\sqrt{5}})^2 + 2 - \sqrt{5}$$

$$= 4 + 3 \cdot \sqrt[3]{(2+\sqrt{5})(4-5)} + 3 \cdot \sqrt[3]{(4-5)(2-\sqrt{5})}$$
$$= 4 - 3 \cdot (\sqrt[3]{2+\sqrt{5}} + \sqrt[3]{2-\sqrt{5}}) = 4 - 3x.$$

Thus, $x^3 = 4 - 3x$. It can be checked that $x = 1$ satisfy the cubic equation

$$x^3 + 3x - 4 = 0,$$

that can be rewritten as $(x - 1)(x^2 + x + 4) = 0$. The quadratic equation $x^2 + x + 4 = 0$ does not have any real roots because its discriminant $1^2 - 4 \cdot 1 \cdot 1 < 0$. Hence, $x = 1$ is the only real solution to the cubic equation $x^3 + 3x - 4 = 0$, and $\sqrt[3]{2+\sqrt{5}} + \sqrt[3]{2-\sqrt{5}} = 1$.
Now, let us simplify the denominator as

$$\sqrt{2} \cdot \sqrt{3-\sqrt{5}} - \sqrt{5} = \sqrt{6-2\sqrt{5}} - \sqrt{5} = \sqrt{1^2 - 2\sqrt{5} + (\sqrt{5})^2} - \sqrt{5}$$

$$= \sqrt{(1-\sqrt{5})^2} - \sqrt{5} = \sqrt{5} - 1 - \sqrt{5} = 1$$

Therefore,

$$\frac{\sqrt[3]{2+\sqrt{5}} + \sqrt[3]{2-\sqrt{5}}}{\sqrt{2} \cdot \sqrt{3-\sqrt{5}} - \sqrt{5}} = 1.$$

With its 546 skyscrapers Hong Kong leads the list of the cities in the world with skyscrapers of 150 meters, though Dubai in the United Arab Emirates is a champion with the most skyscrapers twice as high.

Shared Islands. Let m be the area of Malaysian Borneo in thousand km^2, then $(m + 342)$ and $(m - 192)$ are the areas of Indonesian Borneo and Brunei in thousand km^2. The total or 100% area of Borneo is $m + (m + 342) + (m - 192) = 3m + 150$ thousand km^2. Hence,

$$\begin{array}{cc} \% & \text{thnd. km}^2 \\ 72.58 & m+342 \\ 100 & A=3m+150 \end{array} \quad \Leftrightarrow 3m+150 = \frac{(m+342)\cdot 100}{72.58},$$

or the area m of Malaysian Borneo is 198 thousand km^2. Thus, the area of Indonesian Borneo is 540 thousand km^2 and of Brunei Borneo is 6 thousand km^2. To find the portion in % of Borneo that belongs to Indonesian Borneo and of Brunei Borneo, let us consider proportions for Brunei Borneo

$$\begin{array}{cc} \% & \text{thnd. km}^2 \\ x & 6 \\ 100 & 744 \end{array} \quad \Leftrightarrow x = \frac{6\cdot 100}{744} = 0.8\%,$$

and for Indonesian Borneo

$$\begin{array}{cc} \% & \text{thnd. km}^2 \\ x & 198 \\ 100 & 744 \end{array} \quad \Leftrightarrow x = \frac{198\cdot 100}{744} = 26.62\%.$$

Hence, Indonesian Borneo, called Kalimantan, takes 26.62% and the area of Brunei has 0.8% of the island.

Feeding Animals. Let a be the amount of hay. Cows need a/x of hay per day. Goats and sheep need a/y and a/z of hay per day respectively. All animals in the farm require $a/x + a/y + a/z$ of hay per day. Since the amount of hay is a, the number of days is

$$\frac{a}{\frac{a}{x}+\frac{a}{y}+\frac{a}{z}} = \frac{a}{a\cdot\frac{xy+xz+yz}{xyz}} = \frac{xyz}{xy+xz+yz}.$$

Healthy Weight. Let a/b be a fraction, $0 < a < b$. Then

$$\frac{a+2}{b+2} = 2 \cdot \frac{a}{b} \Rightarrow ab + 2b = 4ab + 4a \Rightarrow 4a = b\,(2-a).$$

If $a > 2$ then $2 - a < 0$ and $b\,(2 - a) < 0$ which leads to contradiction. Namely, the left side $4a$ of the equality is positive, but its right side is negative. Impossible!

If $a = 2$ then $2 - a = 0$ and $b\,(2 - a) = 0$ that cannot be equal to $4a = 4$ under any value of b.

Only $a = 1$ is left. Then $b = 4$.

Indeed, $\frac{1}{4} = \frac{1+2}{4+2} = \frac{3}{6} = \frac{1}{2} = 2 \cdot \frac{1}{4}$.

The BMI of 25 kg/m^2 is healthy.

Sources

en.wikipedia.org/wiki/List_of_countries_with_multiple_capitals

www.worldatlas.com/articles/islands-that-are-shared-by-more-than-one-country.html

www.britannica.com/place/Borneo-island-Pacific-Ocean/History

www.britannica.com/topic/divided-nation

www.history.com/news/north-south-korea-divided-reasons-facts

en.wikipedia.org/wiki/Vietnam

www.britannica.com/topic/divided-nation

en.wikipedia.org/wiki/1954_Geneva_Conference

en.wikipedia.org/wiki/Soviet_Union

www.bbc.com/news/world-europe-17858981

en.wikipedia.org/wiki/Yugoslavia

kids.kiddle.co/Yugoslavia

en.wikipedia.org/wiki/Hay#:~:text

www.scoopwhoop.com/life/interesting-facts-national-anthems-different-countries/

en.wikipedia.org/wiki/Marcha_Real#:~:text=The%20Marcha%20Real%20

www.news24.com/News24/6-facts-about-national-anthems-20140704

en.wikipedia.org/wiki/List_of_aspect_ratios_of_national_flags

en.wikipedia.org/wiki/List_of_countries_with_the_most_skyscrapers

en.wikipedia.org/wiki/List_of_cities_with_the_most_skyscrapers

worldpopulationreview.com/country-rankings/obesity-rates-by-country

1.3 Facts in Factors

Oldest European Countries

All countries value their long history, exceptional culture, and unique heritage. Located on the northeastern side of the Apennine Mountains in Europe, a landlocked country, the Republic of San Marino, claims to be the oldest European country and the oldest constitutional republic as it was formed in 301 AD. Bulgaria is the only European country that has not changed its name since it was first established by the Proto-Bulgarian people who settled south of the Danube River in the 7th century. Founded in 1143, Portugal has the oldest unchanged borders, while France has had its borders changed several times since 843, but its current European border is the same as in medieval times.

Problem. Number N can reveal various information about the oldest countries in Europe. The sum of its digits tells the area of France in ten thousand square miles, its first two digits disclose the area of Portugal in thousand square miles, while the population in San Marino in thousand people is expressed by the thousands and tens of N put side to side and the population in Bulgaria in million people is a decimal fraction with the second to the left digit of N as its integer and the second to the right digit of N as its decimal. Find the magic number N if it is the lowest number that has the remainder of 1 after division by 2, the remainder of 2 after division by 3, the remainder of 3 after division by 4, the remainder of 4 after division by 5, the remainder of $n -$ 1 after division by n, and so on until N stays the same after division by three consequent numbers that follow this pattern.

Answer. 360,359

The Grand Bazaar

The Grand Bazaar inside the Walled City of Istanbul in Turkey is the largest and oldest covered market in the world and the most-visited tourist attraction in the country. *Bazaar* in Turkish is *Covered Market*. Its construction started after the Ottomans conquered Constantinople in 1455. The Grand Bazaar has experienced devastating fires and earthquakes as well as numerous renovations and extensions throughout its long history. Four gates lead to its over 4000 shops filled with colorful diverse merchandise offering an iconic shopping experience for locals and tourists.

It is interesting to note the lack of advertising because the most precious items are already on display. Due to the high security and low crime, each stall is covered only with drapes for a night.

Problem. Find the number N of streets on which the Grand Bazaar in Istanbul is located if this number N is the lowest number that has the remainder of 1 when divided by

- a number n,
- all numbers greater than 1 but less than n,
- the next consequent number $n + 1$.

Answer. 61

The Stan Countries

Several adjacent countries in Central Asia are called the *Stan countries* because their names end up in *stan*, meaning *land of* in Persian. The beginning of their names emphasizes their heritage. Indeed, the first part of *Kyrgyzstan* comes from the Turkic word *kyrgyz*, meaning *forty* and, thus, *Kyrgyzstan* can be seen as *the land of 40* or *we are 40*, mentioning the 40 clans united by the founder of the Kyrgyz Khanate, legendary Manas, in the 9th century.

The Stan countries are culturally and historically linked, have similarities in language, customs, and religion, though their establishment as countries are varied: some of them gaining independence after the Soviet Union collapsed in 1991, others were under different rulers. If united, the Stan countries would have occupied a territory comparable to the US with a much higher population.

Problem. The number of the Stan countries coincides with the first equal remainders obtained after division a certain number by 8 or 10. In addition, this number has the remainder of 1 after division by 2 and the remainder of 2 after division by 3. How many countries are called the Stan countries? Name those that you know.

Answer. 7

Stonehenge

Stonehenge in England is one of the most famous prehistoric landmarks. It has gained legal protection for having cultural and scientific importance since becoming a UNESCO World Heritage Site in 1986. It took over 1000 years to build this unique huge stone circle on Salisbury Plain in Wiltshire during the late Neolithic Age over 5000 years ago. As for other stone monuments, its purpose, the way it was built, and people who built it are a mystery, though different theories and legends exist. One of them tells that Stonehenge stones were moved by a wizard. It is believed that Stonehenge was used for religious purposes and as an astronomical calendar as well as a place of healing and funeral ceremonies.

Problem. How many stones make a stone circle in Stonehenge if this number is the least number that when divided by 6 leaves a remainder of 2, divided by 14 leaves a remainder of 10, and divided by 21 leaves a remainder of 17?

Answer. 38

The Mayan Calendar

The ancient Mayan calendar dated from earlier than the 1st century BCE is tremendously accurate though the most complex and precise among all known ancient and modern calendar systems. It simultaneously used 20-day months and two calendar years; the *Sacred Round* or *tzolkin* and *Vague Year* or *haab* coincided in a certain interval of time, called a *bundle*. The end of a bundle was feared with a belief that the world might come to an end and the sky might fall. No one knows the origin of such an unusual calendar though many hypotheses exist. The ancient Mayan calendar is still used in southern Mexico.

Problem. Using the minimum number of hints, find the time interval when the Sacred Round and the Vague Year of the ancient Mayan calendar coincide. Are all hints needed?

A. The number of solar years is the smallest number that has a remainder of 2 after division by 10, a remainder of 8 after division by 11, and a remainder of 4 after division by 16.
B. The Mayan calendar has 20 days in a month.
C. The Mayan calendar has two calendar years, the 260-day Sacred Round and the 365-day Vague Year.
D. The Sacred Round consists of two smaller cycles.

Answer. 52, no

National Parks

Each nation preserves its national parks. Australia followed by Thailand start the list of countries with the most numbers of parks, while national parks comprise the largest area in Canada and Australia. Few countries can compete with Canada or Australia for their size, though some of them generously provide a large portion of their land to national parks. Zambia is the top country in the world as 32% of its land is given to national parks.

Problem. If the number of national parks in Australia is divided by the number of national parks in Canada, then the quotient will be 14 and the remainder will be 27. If the number of national parks in Australia is divided by the number of national parks in Zambia, then the quotient will be 24 and the remainder will be 13. If the number of national parks in Australia is divided by the sum of numbers of national parks in both countries, then the quotient will be 9 and the remainder will be 10. How many national parks are in Australia, Canada, and Zambia?

Answer. 685, 47, 28

Treasures of the Sea

The ocean keeps sunken ships and vessels and does not rush to allow its treasure to be reclaimed.

Following Columbus' discovery in 1492, the riches of the New World colonies helped Spain to become the most powerful European nation. Vessels filled with gold, gems, copper, silver, and tobacco were shipped to Spain. Not all voyages were successful. The Spanish galleons *Nuestra Senora de Atocha* and her sister ship *Santa Margarita* are in the list of tragic records. These heavily laden ships left Havana for Spain but experienced a horrible hurricane and found their final resting place at the bottom of the ocean just 40 miles off the Florida Keys in September 1622. Precious metals, raw emeralds, and high-quality gems, among other goods had been waiting for centuries to be discovered by the late treasure hunter Mel Fisher in 1985.

Problem. How many tons of gold and silver bars and kilograms of raw emeralds were found at the sunken Spanish galleons *Nuestra Senora de Atocha* and *Santa Margarita* if the product of these two numbers is 1536 and their Least Common Multiple (LCM) is 96? None of the weight numbers coincides with the given product or LCM. The number that represents the weight in tons is greater than the number that stands for the weight in kilograms.

Answer. 48 t, 32 kg

Voting and Drinking Ages

Many teenagers dream about the day when they reach a legal age and gain freedom from their parenting control. Then they become fully responsible for their actions. Based on numerous criteria, this age is different around the world. Voting and drinking ages vary from country to country. The voting age is 16 in Argentina, Austria, Brazil, Ecuador, Nicaragua, and Wales, while 21 in Oman, Tonga, Singapore, and Samoa. It is permitted to drink in Mali and the Central African Republic if you turn 15, and you must wait for your 25th birthday in Eritrea before raising your first legal toast, while you can never do this in Maldives, Sudan, Somalia, or Kuwait where drinking is illegal. On the other side, Burkina Faso, Djibouti, Togo, and Vanuatu are among countries without any legal minimum drinking age.

Problem. Find the voting and drinking age in most countries if this number is the lowest number that has a remainder of 2 after division by 4, and a remainder of 3 after division by 5.

Answer. 18

Movies with Numeric Titles

Twelfth Night by William Shakespeare, *The Three Musketeers* by Alexandre Dumas, *A Tale of Two Cities* by Charles Dickens, *The Sign of Four* by Arthur Conan Doyle, *1984* by George Orwell, *Twenty Thousand Leagues Under the Sea* by Jules Verne, *Three Men in a Boat* by Jerome K. Jerome, *The 101 Dalmatians* by Dodie Smith, *11/22/63* by Stephen King are just a few examples from thousands of books, plays, movies that have numbers in their titles.

A 2022 South Korean television series directed by Jung Ji-hyun and starring Kim Tae-ri, Nam Joo-hyuk, Kim Ji-yeon, Choi Hyun-wook and Lee Joo-myung is not an exception. The show that follows romantic lives of five characters during the 24-year period became a commercial hit and the highest-rated drama in Korean TV history.

Problem. The title of a successful 2022 South Korean television series is presented as a number. This number is the lowest one that leaves the remainder of 1 after division by any integer from 2 to 10. What is the tile of the 2022 South Korean television series?

Answer. 2521

Penicillin and Bubble Gum

The discovery of penicillin is one of the greatest scientific breakthroughs made by accident. Indeed, studying the staphylococcus bacteria, Scottish microbiologist Sir Alexander Fleming (1881–1955) added some of the bacteria to Petri dishes and left for a two-week vacation. He did not wash properly Petri dishes. Upon his return, Sir Fleming noticed that a mold had started to grow but the staph bacteria had deterred. The modern miracle, the first antibiotic in the world, that he called *Penicillin*, was born and is now used to treat everything from acne to pneumonia.

Bubble gum was also invented that year. Although American accountant Walter E. Diemer (1904 –1998) claimed that his discovery was accidental, he tested numerous recipes for a bubble gum before suggesting it at the age of 23.

Problem. Penicillin and bubble gum were discovered, and an electric razor was patented in the year 3610 given in octal numeral system. Find its decimal and binary presentations.

Remark. The octal numeral system or the base-8 numeral system uses the digits 0 to 7. It is widely employed in computers, aviation, and other applications.

Answer. 1928, 11110001000

Solutions

Oldest European Countries. A number N that has a remainder of $n - 1$ after division by n can be written as $N = nk - 1$, where k is any integer. If there are two divisors n and m that are co-primes, then $N = nmk - 1$, otherwise $N = \text{LCM}(nm) \cdot k - 1$, where LCM($nm$) is the Least Common Multiple of n and m, and so on. Hence, considering factorization of composite numbers, we can find out that the first three consequent numbers that have unchanged N are 13, 14, and 15. Indeed, these and all numbers less than these numbers are included in the product $2 \cdot 3 \cdot 2 \cdot 5 \cdot 7 \cdot 2 \cdot 3 \cdot 11 \cdot 13 = 360360$. Then, $N = 360359 = 360360 - 1$. The sum of its digits $3 + 6 + 0 + 3 + 5 + 9 = 26$ tells the area of France in ten thousand square miles, its first two digits 36 reveal the area of Portugal in thousand square miles, while the population of San Marino is 36 thousand people, and 6.5 million people live in Bulgaria.

The area of France is approximately 260,000 square miles, while Portugal takes around 36,000 square miles. 35 thousand people live in San Marino, while 6.5 million people call Bulgaria their home.

The Grand Bazaar. 1st way. The number N divided by another number n with the remainder of 1 can be presented as $N = kn + 1$, where k is any natural number. Thus, the lowest number divided by two consequent numbers 2 and 3 with the remainder 1 is $2 \cdot 3 + 1 = 7$, by 2, 3, and 4, and 5 with the remainder 1 is $2 \cdot 3 \cdot 2 + 1 = 13$. Note that, instead of adding multiplication by 4, multiplication by another 2 is added as 4 is presented by its prime factorization 2^2 or $2 \cdot 2$ and one of two 2s has been already in the product, that is, the product should contain all factors of all considered numbers. The next number that has a remainder of 1 after division by 2, 3, and 4, and 5 is $2 \cdot 3 \cdot 2 \cdot 5 + 1 = 61$. Since the prime factorization of 6 is $2 \cdot 3$ and 2 and 3 are already in the product $2 \cdot 3 \cdot 2 \cdot 5$, then the number 61 can also be divided by 6 with a remainder of 1.

2nd way. For a number to have a remainder of 1 after division by several numbers, the number should be their Least Common Multiple (LCM) plus 1. Let us consider one-digit numbers and find their LCM:

Numbers	LCM	LCM + 1
2, 3	$6 = 2{\cdot}3$	7
2, 3, 4	$12 = 2{\cdot}3{\cdot}2$	13
2, 3, 4, 5	$60 = 2{\cdot}3{\cdot}2{\cdot}5$	61
2, 3, 4, 5, 6	$60 = 2{\cdot}3{\cdot}2{\cdot}5$	61

Over 61 covered streets make up the Grand Bazaar in Istanbul, Turkey.

The Stan Countries. A number that has a remainder of 1 after division by 2 and a remainder of 2 after division by 3 can be presented as $2{\cdot}3{\cdot}n - 1$, where n is any natural number. There are several ways to continue solving the problem. Let us choose the most visual one and prepare a table that will also reveal some interesting patterns:

n	$2{\cdot}3{\cdot}n$	$2{\cdot}3{\cdot}n - 1$	Remainder after division by 8	Remainder after division by 10
1	6	5	3	1
2	12	11	1	7
3	18	17	7	3
4	24	23	5	9
5	30	29	3	5
6	36	35	1	1
7	42	41	7	7
8	48	47		

The first remainder that appears in both rows is 1, but it does not go along with the condition that there are several Stan countries. The next suitable remainder is 7.

It is interesting to note the repetition of remainders 3, 1, 7, 5 and 1, 7, 3, 9, 5 in the last two columns.

Afghanistan, Kazakhstan, Kyrgyzstan, Pakistan, Tajikistan, Turkmenistan, and Uzbekistan are the Stan countries. Five of them used to be republics of the former Soviet Union.

Stonehenge. 1st way. Let us check the differences between divisors and related remainders: $6 - 2 = 4$, $14 - 10 = 4$, $21 - 17 = 4$. Fortunately, they are the same! Finding the Least Common Multiple (LCM) of $6 = 2{\cdot}3$, $14 = 2{\cdot}7$, and $21 = 3{\cdot}7$ as $2{\cdot}3{\cdot}7 = 42$ and subtracting 4, we get $42 - 4 = 38$.

2nd way. Let N be the number of stones. Then $(N - 2)/6$, $(N - 10)/14$, and $(N - 17)/21$ are integers and

$$\frac{N-2}{6}+1=\frac{N+4}{6}, \frac{N-10}{14}+1=\frac{N+4}{14}, \text{and} \frac{N-17}{21}+1=\frac{N+4}{21},$$

are also integers. It means that $N + 4$ is the LCM of $6 = 2 \cdot 3$, $14 = 2 \cdot 7$, and $21 = 3 \cdot 7$. So, $N = 42 - 4 = 38$.

Stonehenge is a stone circle made up of 38 stones on Salisbury Plain in Wiltshire.

The Mayan Calendar. Let N represent the number of years after which the Sacred Round and the Vague Year meet.

1st way. Following Hint A, N can be presented as $10k + 2$, as $11n+8$, or as $16m+4$, where k, n, and m are any positive integers. Hence, $10k + 2 = 11n+8 = 16m+4$, from which follows

$$k=\frac{1+8m}{5} \quad \text{and} \quad n=\frac{16m-4}{11}.$$

In order for $(1 + 8m)$ to be divisible by 5, $8m$ should end up in 4 that can occur if m is any positive integer ended in 3 or 8. Let us test the lowest value of $m = 3$, then $n = 4$, and $N = 52$.

2nd way. Following Hint C, let us find the Least Common Multiple (LCM) of 260 and 365 using prime factorization. Presenting 260 as $2^2 \cdot 5 \cdot 13$ and 365 as $5 \cdot 73$, we get LCM $(260, 365) = 2^2 \cdot 5 \cdot 13 \cdot 73 = 18{,}980$ days or $18{,}980 / 365 = 52$ years.

3rd way. Instead of finding LCM (260, 365), we can simply multiply factors of 260 not inclusive in 365, as $2^2 \cdot 13 = 52$.

4th way. Following Hint C, let us find the Greatest Common Divisor of 260 and 365 using the Euclidean Algorithm as

$$365 = 260 \cdot 1 + 105$$

$$260 = 105 \cdot 2 + 50$$

$$105 = 50 \cdot 2 + 5$$

$$50 = 5 \cdot 10$$

to get 5 as their Greatest Common Divisor. Then Least Common Multiple LCM (260, 365) is

$$\frac{365 \cdot 260}{5} = 18980,$$

which gives $18980 / 365 = 52$ years.

The Sacred Round and the Vague Year coincide every 52 years. It is interesting to note that our year has 52 weeks.

National Parks. The number a of national parks in Australia can be presented via the numbers of national parks in Canada c and Zambia z as $14c + 27 = a$ and $24z + 13 = a$, from which follows

$$c = \frac{a-27}{14} \text{ and } z = \frac{a-13}{24}.$$

Substituting these expressions to the equation $9(z + c) + 10 = a$, we get

$$9\left(\frac{a-27}{14} + \frac{a-13}{24}\right) + 10 = a, \text{ that leads to } a = 685, z = 28, c = 47.$$

There are 685 national parks in Australia, 47 in Canada, and 28 in Zambia.

Treasures of the Sea. Let t and k be amounts of gold and silver bars in tons and raw emeralds in kilograms. The product of the two numbers $tk = 1536$ and their LCM$(t, k) = 96$, then GCF$(t, k) = 1536 / 96 = 16$ and t and k can be presented as $t = 16n$ and $k = 16m$, where n and m are any positive integers such that $nm = 96/16 = 6$. The following options for the set of positive integers (n, m) are (1, 6), (6, 1), (2, 3), (3, 2). They produce the sets of (t, k) as (16, 96), (96, 16), (32, 48), (48, 32). Only the last set $(t, k) = (48, 32)$ satisfies the last condition of the problem.

A total of 48 tons of gold and silver bars and 32 kg of raw emeralds that sank with the Spanish galleons in 1622 were recovered in 1985.

Voting and Drinking Ages. 1st way. Following the problem, the age N can be presented as $N = 4n + 2$ or as $N = 5k + 3$, where n and k are any positive integers, Hence, $4n + 2 = 5k + 3$, from which follows

$$k = \frac{4n-1}{5}.$$

In order for an odd $(4n - 1)$ to be divisible by 5, it should end up in 5. Then n can be 4, 9, 14, 19, or any other number that ends at 4 or 9. Checking the lowest $n = 4$, we get $k = 3$. Thus $N = 18$.

2nd way. Let us check the difference between divisors and related remainders: $4 - 2 = 2$, $5 - 3 = 2$. They are the same! Subtracting 2 from the Least Common Multiple (LCM) of 4 and 5, that is $4 \cdot 5 = 20$, we get $20 - 2 = 18$.

Age 18 is considered the legal adult age with the right to vote. It is allowed to drink at the age of 18 in 64% of the world's nations.

Movies with Numeric Titles. Let us perform a prime factorization of each number from 2 to 10, multiply the minimum number of prime factors that compose these numbers, and add 1 to get

$$1 \cdot 2^3 \cdot 3^2 \cdot 5 \cdot 7 + 1 = 2521.$$

Twenty-Five Twenty-One is the title of a 2022 South Korean television series.

Penicillin and Bubble Gum. The octal or the base-8 numeral system number uses the digits from 0 to 7. Following the formula

$$N = a_n b^n + a_{n-1} b^{n-1} + a_{n-2} b^{n-2} + \ldots + a_1 b^1 + a_0,$$

where b is the base, a_k is a k-positioned digit, $k = 0, 1, \ldots, n$, that transfers numbers in the base-n numeral system to the decimal system, we get

$$3 \cdot 8^3 + 6 \cdot 8^2 + 1 \cdot 8 + 0 = 1928.$$

Presenting 1928 as a sum of powers of 2

$$1928 = 1024 + 512 + 256 + 128 + 8 = 2^{10} + 2^9 + 2^8 + 2^7 + 2^3$$

$$= 1 \cdot 2^{10} + 1 \cdot 2^9 + 1 \cdot 2^8 + 1 \cdot 2^7 + 0 \cdot 2^6 + 0 \cdot 2^5 + 0 \cdot 2^4 + 1 \cdot 2^3 + 0 \cdot 2^2 + 0 \cdot 2^1 + 0 \cdot 2^0,$$

leads to its binary equivalent 11110001000.

Scottish physician and microbiologist Sir Alexander Fleming discovered penicillin, American accountant Walter E. Diemer invented bubble gum, and American inventor and entrepreneur Colonel Jacob Schick patented an electric razor and started the Schick Dry Shaver, Inc. company in 1928.

Sources

www.iexplore.com/experiences/cultural-experiences/Exciting-World-Bazaars-Street-Markets

en.wikipedia.org/wiki/Grand_Bazaar,_Istanbul

en.wikipedia.org/wiki/Portugal

en.wikipedia.org/wiki/France

www.civitatis.com/blog/en/6-fun-facts-world/, en.wikipedia.org/wiki/San_Marino,

worldpopulationreview.com/country-rankings/oldest-country-in-europe

www.worldatlas.com/places/all-about-the-stan-countries.html

www.funkidslive.com/learn/top-10-facts/top-10-facts-about-stonehenge/

www.historymuseum.ca/cmc/exhibitions/civil/maya

safarisafricana.com/which-country-has-the-most-national-parks/

www.melfisher.com/library/atochamargstory.asp, en.wikipedia.org/wiki/Nuestra_Se%C3%B1ora_de_Atocha

en.wikipedia.org/wiki/Voting_age

vinepair.com/articles/legal-drinking-age-world-map/

en.wikipedia.org/wiki/Twenty-Five_Twenty-One

www.mynewlab.com/blog/accidental-scientific-discoveries-and-breakthroughs/

1.4 Sequential History

National Anthems around the Globe

The national anthem is a patriotic or folk song or musical tune that uniquely represents a country. The exotic island country of samurai and ninjas, Japan, has the shortest national anthems, *Kimi Ga Yo*, taking its words from the *Kokin Wakashū* first published in 905 and set to music in 1880. This national anthem has the oldest lyrics among all other national anthems. Sadly, it is still performed in an archaic language hardly understandable by the younger generation. The Dutch claim to have the oldest melody for its national anthem, *Wilhelmus*, composed in 1574, while the British national anthem *God Save the King* is the oldest publicly recognized national anthem as it was performed in London in 1745.

The country of Olympic Games and antient mythology, Greece, is proud of its longest national anthem in the world. It takes nearly one hour to sing the full version of the *Hymn to Liberty* or *Hymn to Freedom*, written by Dionysios Solomos in 1823 with added musical tunes by Nikolaos Mantzaros in 1865. It shortened version was adopted in 1865. The Ugandan national anthem *Oh Uganda, Land of Beauty*, is musically one of the shortest national anthems in the world. Its lyric and music by George Wilberforce Kakoma were adopted as the national anthem after the country gained independence from the United Kingdom in 1962. Composed by Francisco Jose Debali and Fernando Quijano to lyrics by Francisco Acuña de Figueroa and accepted in 1848 as the national anthem of Uruguay, the *Himno Nacional de Uruguay*, or *Orientales, la Patria o la Tumba*, is now the longest national anthem performed. Therefore, only the first verse and chorus are sung on most occasions.

Problem. The number of lines in the Japanese national anthem is the second term of an arithmetic sequence with 60 as the sum of its first six terms and 120 as the sum of the following four terms. The number of bars in the Ugandan anthem music coincides with the third term of the arithmetic sequence while the number of music bars in the national anthem of Uruguay is one more than the 27th term. The number of stanzas in the Greek national anthem is two more than the 40th term of the arithmetic sequence. Find these numbers.

Answer. 4, 8, 105, 158

Populations of the Seven Continents

Population density rates describe the concentration of people per unit area in geographic locations. Population density data is widely used to make predictions, estimates, and address other issues, such as evaluation of various relationships and allocation of production.

Problem. Population densities in people/km^2 at the continents of Asia, Africa, Europe, South America, Australia and Oceania, and Antarctica, are respectively the 22nd, 10th, 8th, 6th, 2nd, and 1st terms of an increasing arithmetic sequence, while the population density in North America is 0.5 more than the average of its 6th and 7th terms. The sum and product of the 3rd and 5th terms of this arithmetic sequence are 30 and 200 correspondingly. Find a population density at the seven continents.

Answer. 105, 45, 35, 28, 25, 5, 0

The Land Down Under

The Commonwealth of Australia is the largest country by area in Oceania and the world's sixth-largest country. It has been inhabited for over 65 millenniums. The European exploration of Australia started with landfall of Dutch ships led by Willem Janszoon that named the western and northern coastlines of the discovered land *New Holland*. The British led by Captain James Cook arrived later and claimed the east coast, called *New South Wales*, for Great Britain. The European population had grown fast, forming a colony in New South Wales. The Union flag was raised at Sydney Cove on January 26, a date that became Australia's national day. Later, the name of the country changed from *New Holland* to *Australia*. The country is also nicknamed *The Land Down Under*, *The Great Outback*, and *The Lucky Country*.

Problem. Find the years when Australia was discovered by English, the Union flag was raised at Sydney Cove, and the continent got its recent name, if these years can be presented as 29th, 32nd, 38th terms of an increasing arithmetic sequence with the first term of 1602. The product of two consequent terms is 3,250,800 and the sum of their squares is 6,501,636. Find the year when Australia was discovered by the Dutch if it is 2 less than the second term of this sequence.

Answer. 1770, 1788, 1824, 1606

From Clasp Lockers to Zippers

You will be surprised to find out that a *clasp locker*, as a *zipper* was originally called, got neither any attention nor a commercial value after being invented and patented by American inventor Elias Howe and later by Whitcomb Judson. Its design was improved and manufacturing machines for the new device were created by American, Australian, and Swiss inventors. Initially used in the production of footwear and tobacco bags, zippers eventually became very popular.

Problem. Find the years when a clasp locker was invented, introduced to the public, and named a zipper, if these years can be presented as the 9th, 16th, and 21st terms of an arithmetic sequence. The sum of its fourth and seventh terms is 3660, and their difference is 18.

Answer. 1851, 1893, 1923

Microwaves

A microwave is one of the most common appliances in our houses and offices. This was not always the case: when the first microwaves appeared, no one thought they could be used for cooking. While developing energy sources for radar equipment and working with magnetrons, American physicist and engineer Percy LeBaron Spencer noticed that a candy bar in his pocket had melted. After testing his accidental discovery on different items, he patented the microwave cooking process. The microwave oven for household use was introduced later.

Problem. One of the 20th-century inventions was the microwave. The last two digits of the year when microwaves were designed form the number that stands for the sum of an infinite geometric alternating sequence. Both the first term and the sum of the sequence have over five factors, but their Greatest Common Divisor (GCD) is 3 and their Least Common Multiple (LCM) is 720.

Answer. 1945

The Deepest Point

The Pacific Ocean is the largest and deepest ocean on the Earth. The Challenger Deep located in the southern end of the Mariana Trench close to the US territorial island of Guam is the deepest point in the ocean and on the Earth. If Mount Everest were placed at the bottom of the Mariana Trench, its peak would still be over 7000 feet below sea level.

Problem. How deep is the Challenger Deep if its depth in miles is the first term of an infinite geometric series with the sum of 9.1? The 26th term of the geometric series is 64 times more than its 29th term.

Answer. 6.825 mi

Countries with Many Borders

What can you say about a country that has many neighboring countries? Probably, the country is big and has a long land border. The answer is *Yes* and *No*. Despite having the world's longest border, Canada borders only the United States. On the other side, Russia and China are the first and third countries in the world by the area. They take a huge part of Asia. Furthermore, Russia is also located in Europe. They share first place if the number of bordering countries is considered. Brazil dominates South America as its largest country. Only Ecuador and Chile do not border with Brazil that takes a leading spot in the list of countries with the most international borders.

Problem. China and Russia followed by Brazil have the greatest number of bordering countries. These numbers are the first and second terms of an infinite geometric sequence. The sum of all terms of the sequence is 49 and the ratio between the 7th and 4th terms is 125 to 343. How many countries do they border with? Can you name nations that share borders with China, Russia, and Brazil?

Answer. 14, 10

Decline of the Olympic Games

The most famous of all sporting festive started in ancient Greece were held every four years in honor of Zeus in a sacred site called Olympia located near the western coast of the Peloponnese peninsula in Greece. All Olympic champions were treated as national heroes. Their names were thoroughly documented, and some records have survived to our day. The Olympic Games in Greece lasted for over a thousand years and continued even when the Roman Empire conquered Greece, though their standards and number of sporting events significantly declined under Roman rulers. Finally, Emperor Theodosius I, a Christian, banned all *pagan* rituals including the Olympic Games.

Problem. Find the year of the last ancient Olympic Games if this year is the sum of all terms of the decreasing geometric sequence with 81 as the ratio between its second and sixth terms, and $88\frac{100}{243}$ as their sum.

Answer. 393

Ancient Olympic Games

The Olympic Games are the world's leading sporting event. A legend says that Heracles who was a son of Zeus (the king of all gods on Mount Olympus) and Alcmene (a mortal princess) founded the Games in honor of Zeus. The exact time of the first games is unknown. Therefore, their count started from the date of the first found records honoring the Olympic champion who won a 192-meter footrace or the *stade*, from which the word *stadium* originated. The games were held every four years in Olympia. Only freeborn males of Greece could participate in the ancient Olympic Games. Married women were even prohibited from attending the events.

Problem. Find the year in BCE of the first Olympic Game if the year is the ninth term of an arithmetic sequence with $x^2 - y^2$, xy, $x^2 + y^2$ as its first three terms and 9506 as its 99th term. All terms of the arithmetic sequence are nonnegative.

Answer. 776 BCE

Modern Olympic Games

The ancient Olympic Games were very popular in Greece for over a thousand years until a Roman emperor banned them. It took the enthusiasm and dedication of Baron Pierre de Coubertin (1863–1937) of France to revive the games 1500 years later. He also introduced the Olympics motto *Citius, altius, fortius* which means *Faster, higher, stronger*. The flag of the Olympic Games has five interlocking rings on a white background. The rings represent five parts of the world and at least one of their blue, yellow, black, green, and red colors are present in any national flag. The first modern Olympic Games were held in Athens, Greece, to commemorate continuation of the ancient games. 280 participants from 12 nations competed in 43 events. The running event followed the 25-mile route made by a Greek soldier who brought victory news from *Marathon* (from which the long-distance *marathon* got its name) to Athens in 490 BC. Amazingly, a Greek won the first modern marathon.

The International Olympic Committee with a headquarters in Switzerland decides on a variety of sports to be included, chooses the cities to host the Games, and makes other major decisions. As in ancient times, the Olympic Games are held every four years and numbered accordingly even if no Games take place as during World War I in 1916, and during World War II in 1940 and 1944.

It is interesting that for about 80 years only amateur athletes were allowed to compete.

Problem. Find the year of the revival of the Olympic Games if this year is six times the 53rd term of an arithmetic sequence with the following properties.

A. x/y, $x + y$, xy are the first three terms of the arithmetic sequence,
B. x is the smallest positive integer that satisfies Properties A, C, and D,
C. y is the largest rational number that satisfies Properties A, B, and D,
D. the arithmetic sequence is increasing,

Answer. 1896

Solutions

National Anthems around the Globe. Let a, a_n, and d be the first term, the n-th term, and the difference of an arithmetic sequence, then the sum of its first six terms is

$$S_6 = \frac{2a+(6-1)d}{2}\cdot 6 = 60,$$

and, because $a_7 = a + 6d$, the sum of the following four terms is

$$S_{7 to 10} = \frac{2(a+6d)+(4-1)d}{2}\cdot 4 = 120.$$

Simplifying two equations above, we get the system

$$\begin{cases} 2a+5d=20 \\ 2a+15d=60 \end{cases}'$$

from which it follows that $a = 0$, $d = 4$. Hence, $a_2 = 4$, $a_3 = 8$, $a_{27} = 104$, $a_{40} = 156$.

Japan is the country with the shortest four-line national anthem *Kimigayo,* translated as *His Imperial Majesty's Reign.* With just eight bars of music, the Ugandan anthem is musically one of the shortest anthems in the world. It is occasionally performed twice in a row to lengthen it. The longest national anthem in terms of duration with 105 bars of music is in Uruguay. With 158 stanzas, Greece has the longest national anthem in the world.

Populations of the Seven Continents. Let a and d be the first term and the difference of the arithmetic sequence, then $a_3 = a + 2d$ and $a_5 = a + 4d$ are its 3rd and 5th terms, and

$$a_3 + a_5 = a + 2d + a + 4d = 2a + 6d = 30, \text{ or } a = 15 - 3d,$$

$$a_3 \cdot a_5 = (a + 2d) \cdot (a + 4d) = (15 - 3d + 2d) \cdot (15 - 3d + 4d) = (15^2 - d^2) = 200,$$
$$\text{or } d = 5, \text{ and } a = 0,$$

Another solution $d = -5$ of $15^2 - d^2 = 200$ is not considered because the arithmetic sequence is increasing. Then the 6th and 7th terms are $a_6 = a + 5d = 25$ and $a_7 = a + 6d = 30$, and their average is $\frac{25+30}{2} = 27.5$.

The population density in Asia is 105 people per km^2, in Africa is 45, in Europe is 35, in North America is 28, in South America is 25, in Australia and Oceania is 5, in Antarctica is 0 people per km^2.

The Land Down Under. Let a, $a + d$, and $d > 0$ are two consequent terms and the difference of an increasing arithmetic sequence, then, following the problem, we get

$$\begin{cases} a(a+d) = 3250800 \\ a^2 + (a+d)^2 = 6501636 \end{cases}.$$

Substituting $a + d = \frac{3250800}{a}$ from the first equation to the second equation, we get

$$a^2 + \left(\frac{3250800}{a}\right)^2 = 6501636,$$

that after a substitution $t = a^2$, leads to a quadratic equation $t^2 - 6501636t + 3250800^2 = 0$ that has two solutions $t = 3240000$ and $t = 3261636$. Hence, $a = 1800$ and $d = 6$, or $a = 1806$ and $d = -6$. The second option does not work because the sequence is increasing.

From the problem, $a_1 = 1602$. Using the formula for the n-th term for an arithmetic sequence $a_n = a_1 + d(n - 1)$, we get $a_2 = 1602 + 6{\cdot}1 = 1608$, $a_{29} = 1602 + 6{\cdot}28 = 1770$, $a_{32} = 1602 + 6{\cdot}31 = 1788$, and $a_{38} = 1602 + 6{\cdot}37 = 1824$.

The first Europeans landed in Australia on February 26, 1606. On April 26, 1770, Captain James Cook sailed along the east coast. Great Britain started the colony of New South Wales on 26 January 26, 1788. The country was named Australia in 1824.

From Clasp Lockers to Zippers. The arithmetic sequence is increasing because the difference between its 4th and 7th terms is positive. Moreover, the difference of the arithmetic sequence $d = 6$ because there are three terms between the 4th and 7th terms meaning that d is added three times to the 4th term to get the 7th term and $18/3 = 6$. Then, $a_4 + a_7 = a_4 + a_4 + 18 = 3660$, or $a_4 = 1821$ and $a_9 = a_4 + 5d = 1851$, $a_{16} = a_4 + 12d = 1893$, and $a_{21} = a_4 + 17d = 1923$

Elias Howe received a patent for an *Automatic, Continuous Clothing Closure* in 1851. Based on this idea, Whitcomb Judson patented a *Clasp Locker* in 1893. The B. F. Goodrich Company named the invention *zipper* in 1923.

Microwaves. Let us present the problem solution in steps.

- Since the greatest common divisor GCD of the first term and the sum of an infinite geometric sequence is 3, they can be presented as $3a$ and $3s$ correspondingly, where a and s are any integers and GCD $(a, s)=1$.
- Then $3as = 720$ or $as = 240$.
- Listing all pair of multiples of 240 as (1, 240), (2, 120), (3, 80), (4, 60), (5, 48), (6, 40), (8, 30), (10, 24), (12, 20), (15, 16), we can see that only (15, 16) works because
  - at least one factor of (1, 240), (2, 120), (3, 80), (5, 48) is not composite,
  - pairs (2, 120), (4, 60), (6, 40), (8, 30), (10, 24), (12, 20) have another common factor.
  - the first set is taken away because it contradicts the condition of the numbers having more than 5 factors,
  - the second set is taken away because it contradicts the condition that two numbers have only one common factor of 3.
- Let us consider the pair (15, 16). There are two options:
  - the first term b of an infinite geometric alternating sequence is 45 and the sum S of all terms is 48, then from the formula for the sum of an infinite geometric sequence
  - $S = \frac{b}{1-r}$ we get $48 = \frac{45}{1-r}$ or $r = \frac{1}{15}$.

  This geometric series has only positive terms.
- $b = 48$ and $S = 45$, then

$$S = 45 = \frac{48}{1-r} \quad \text{or} \quad r = -\frac{1}{15},$$

that produces an infinite geometric alternating sequence. Therefore, $S = 45$.

American physicist, engineer, and inventor Percy Spencer invented a microwave in 1945.

The Deepest Point. Let a and q be the first term and the common ratio of a geometric sequence. Using the formula for the n-th term of a geometric sequence $a_n = a \cdot q^{n-1}$, we get $a_{29} = a \cdot q^{28}$ and $a_{26} = a \cdot q^{25}$, that leads to $q = 1/4$. From the general formula for the sum of infinite geometric sequence

$$S = \frac{a}{1-q} \text{ follows } S = \frac{a}{1-1/4} = 9.1 \text{that leads to } a = 9.1 \cdot \frac{3}{4} = 6.825.$$

Challenger Deep in the Mariana Trench is the Earth's deepest point reaching a depth of 6.825 miles, 10,983 meters, or 36,036 feet.

Countries with Many Borders. Let the first term and the common ratio of a geometric sequence be a and r. Then its fourth and seventh terms are $a_4 = ar^3$ and $a_7 = ar^6$. Following the problem, we get the ratio

$$\frac{ar^6}{ar^3} = r^3 = \frac{125}{343} \text{ or } r = \frac{5}{7}.$$

Using the formula for the sum of an infinite geometric sequence $S = \frac{a}{1-r}$, we get $\frac{a}{1-\frac{5}{7}} = 49$ or $\frac{7a}{2} = 49$ from which $a_1 = a = 14$ and $a_2 = 10$.

Afghanistan, Bhutan, India, Kazakhstan, Kyrgyzstan, Laos, Mongolia, Myanmar, Nepal, North Korea, Pakistan, Russia, Tajikistan, and Vietnam are 14 bordering countries of China.

Russia's 14 bordering countries are Azerbaijan, Belarus, China, Estonia, Finland, Georgia, Kazakhstan, Latvia, Lithuania, Mongolia, North Korea, Norway, Poland, and Ukraine.

Ten countries, Argentina, Bolivia, Colombia, France (French Guiana), Guyana, Paraguay, Peru, Suriname, Uruguay, and Venezuela, share their borders with Brazil.

Decline of the Olympic Games. Let the first term and the common ratio of a geometric sequence be a and r. Then its second and fifth terms are $a_2 = ar$ and $a_6 = ar^5$. Following the problem, we can present the ratio

$$\frac{ar}{ar^5} = \frac{1}{r^4} = 81 \text{ or } r = \frac{1}{3}.$$

From their sum

$$ar + ar^5 = \frac{a}{3} + \frac{a}{3^5} = \frac{a(3^4+1)}{3^5} = 88\frac{100}{243} = \frac{21484}{243} \text{ follows } a = 262.$$

Using the formula for the sum of an infinite geometric sequence $S = \frac{a}{1-r}$, we get $\frac{262}{1-\frac{1}{3}} = 393.$

The ancient Olympic Games stopped in 393.

Ancient Olympic Games. Since $x^2 - y^2$, xy, $x^2 + y^2$ are the first consequent terms of an arithmetic sequence, then the differences between the third and the second terms and between the second and the first terms are the same,

$$x^2 + y^2 - xy = xy - x^2 + y^2$$

that leads to $2x^2 = 2xy$, and to either $x = 0$ or $x = y$.

If $x = 0$, then the first three terms of the arithmetic sequence become $-y^2$, 0, y^2. The first term is negative. Thus, $x = 0$ does not work.

If $x = y$, then the first three terms of the arithmetic sequence are 0, x^2, $2x^2$, the difference is x^2, and all other terms are presented as $a_n = (n - 1)x^2$, $n \geq 1$. The 99th term is 9506. From $(99 - 1)x^2 = 9506$ follows $x^2 = 97$ and the 9th term $(9 - 1)\cdot 97 = 776$.

According to the known written records, the first Olympic Games took place in Greece in 776 BCE.

Modern Olympic Games. Let us follow the steps presented below.

- From the first term x/y follows $y \neq 0$.
- If $x = 0$, then the first three terms become 0, y, 0 that cannot be the first three terms of an increasing arithmetic sequence. Thus, $x \neq 0$.
- Since x/y, $x + y$, xy are the first consequent terms of an arithmetic sequence, the differences between the third and the second terms and between the second and the first terms are the same,

$$x + y - \frac{x}{y} = xy - (x + y).$$

- Multiplying both sides of the last statement by $y \neq 0$ leads to

$$xy + y^2 - x = xy^2 - xy - y^2 \text{ or } y^2(x - 2) + x(1 - 2y) = 0.$$

- Dividing both sides of the last expression by $xy^2 \neq 0$ and simplifying, we get

$$\frac{1}{y^2} - \frac{2}{y} + \left(1 - \frac{2}{x}\right) = 0$$

that after the substitution $z = 1/y$ becomes a quadratic equation

$$z^2 - 2z + \left(1 - \frac{2}{x}\right) = 0 \text{ with the solution. } z = 1 \pm \sqrt{1 - \left(1 - \frac{2}{x}\right)} \text{ or } z = 1 \pm \sqrt{\frac{2}{x}}.$$

- In order for y to be a rational number, x should be $2a^2$, where a is any positive rational number. Then $z = 1 \pm \frac{1}{a}$, that with the back substitution gives $\frac{1}{y} = \frac{a \pm 1}{a}$ or $y = \frac{a}{a \pm 1}$.

- Let us consider $y = \frac{a}{a+1}$. The lowest integer is $a = 1$. Sadly, $a = 1$ does not satisfy Property D. Indeed, then $x = 2$, $y = ½$ and the first three terms become 4, 3/2, and 1 that decrease.

- The second option $y = \frac{a}{a-1}$ together with $x = 2a^2$ lead to the first three terms $2a(a-1)$, $2a^2 + \frac{a}{a-1}$, $\frac{2a^3}{a-1}$. The smallest integer $a = 2$ gives the largest $y = 2$, and, therefore, satisfies all other Properties A, B, C, and D, producing the first three terms 4, 10, 16, and the formula for the n-th term $a_n = 4 + 6(n-1)$. The 53rd term of the arithmetic sequence is $a_{53} = 4 + 6(53-1) = 316$, which after multiplying by 6 gives 1896.

The first modern Olympic Games were held in 1896 in Athens, Greece.

Sources

kateswanderlust.com/nicknames-of-australia/

www.worldometers.info/geography/7-continents/

en.wikipedia.org/wiki/Australia

en.wikipedia.org/wiki/Terra_nullius

www.ducksters.com/geography/country/australia_history_timeline.php

www.nationalgeographic.org/activity/mariana-trench-deepest-place-earth/

en.wikipedia.org/wiki/Mariana_Trench

www.primaryhomeworkhelp.co.uk/war/inventions.html

www.legalzoom.com/articles/top-5-accidental-inventions-discoveries

www.mynewlab.com/blog/accidental-scientific-discoveries-and-breakthroughs/

www.cbc.ca/news/science/zipper-anniversary

en.wikipedia.org/wiki/Zipper

www.history.com/topics/sports/olympic-games

kids.britannica.com/kids/article/Olympic-Games/353563

www.thoughtco.com/countries-with-most-neighbors

www.worldatlas.com/articles/countries-bordering-the-highest-number-of-other-countries

1.5 Unions of Unions

The United Kingdom

The British Isles in the northwestern coast of Europe consist of two main islands, Great Britain and Ireland, and numerous smaller islands. The tribes and later countries on the British Isles have had an incredibly rich history witnessing unions and separations of its countries.

Several kingdoms united and created England in 927 to repeal the Danish, the Germans, and other foreign invaders. Wales became part of the Kingdom of Great Britain in 1707. The Kingdom of Ireland joined the Kingdom of Great Britain in 1801 establishing the United Kingdom of Great Britain and Ireland. Most of Ireland seceded from the UK in 1922 forming the Republic of Ireland.

Problem. Great Britain and Ireland are the two largest islands of the British Isles. England, Scotland, and Wales are located on the island of Great Britain. The Republic of Ireland and Northern Ireland are located on the island of Ireland. The UK is the union of four independent countries, England, Scotland, Wales, and Northern Ireland.

Depict the British Isles, Great Britain, Ireland, the United Kingdom, England, Northern Ireland, Scotland, and Wales and their connections using a Venn diagram.

Answer. See the solution

The United Nations

The United Nations (UN) is the largest and most famous international organization that aims to maintain international peace, cooperation, security, and relations and address various global issues. The organization was named by USA President Franklin D. Roosevelt. The UN was established after World War II during a meeting in San Francisco, USA. The UN membership has grown fast from its original 51 states to the current 193 sovereign states. Several official languages have been selected for the UN meetings and correspondence. The UN headquarter is in New York City (USA) with main offices in Geneva (Switzerland), Nairobi (Kenya), Vienna (Austria), and The Hague (Netherlands). The Hague is also home to the International Court of Justice. World Bank Group, the World Health Organization, the World Food Program, UNESCO, and UNICEF are among 36 UN specialized agencies and programs.

Problem. What year was the UN established if the year is the product of the sum of all elements in the set A and the sum of all elements in the set C? How many permanent members of the Security Council were upon ratification of the Charter if this number of countries is the sum of all elements of the set C? Name these counties. How many official languages does the organization have if this number is the sum of all elements in $\overline{A} \cap B \cap \overline{C}$? List UN official languages that you know.

Use the following information to address the questions above.

A. $A \cup B = \{5, 6, 45, 56, 68, 69, 72, 79\}$.
B. $A \cap \overline{B} = \{56, 68, 72\}$.
C. $A \cap B = \{45, 69, 79\}$.
D. $B \cap \overline{C} = \{6, 45, 69, 72\}$.
E. $A \cup B \cup C = \{5, 6, 45, 56, 68, 69, 72, 79\}$

Answer. 1945, 5, 6

Continental Unions

Geographical, political, and economic unions of countries have been formed to achieve solidarity between them and promote their political, social, economic, educational, and cultural integration and cooperation.

The Association of Southeast Asian Nations (ASEAN) formed on July 31, 1961 unites Southeast Asian states and aims to promote economic growth, peace, and stability. The ASEAN Secretariat is in the capital of Indonesia, Jakarta. English is the ASEAN official language though ten other languages are in use. The ASEAN Motto is *One Vision, One Identity, One Community.*

The European Union (EU) is a political and economic union of states primarily located in Europe. Established on November 1, 1993, the EU allows the free movement of people, goods, and services within the union. Only 19 EU members use the euro. Brussels, the capital of Belgium, is the de facto EU capital. English, French, and German are among the three most common languages used in the EU. The EU motto is *United in Diversity*.

The African Union (AU) was announced in the Sirte Declaration in Sirte, Libya, on September 9, 1999. The AU Commission is in the capital of Ethiopia, Addis Ababa. Arabic, English, French, Portuguese, Spanish, and Kiswahili are the AU primary languages. Its motto is *A United and Strong Africa*.

Problem. How many countries are in the ASEAN, AU, and EU, if the number of countries in

- the African Union is the product of two largest elements of the $\bar{A}$ or the product of the middle term and the sum of the smallest and largest elements of $\bar{B}$,
- the Association of Southeast Asian Nations is the product of all elements of $\bar{A} \cap \bar{B}$,
- the European Union is the sum of all elements of $\overline{A \cap B}$.
- $A = \{1, 3, 4, 9\}$ and $B = \{1, 3, 4, 11\}$ are two events of the sample space $S = \{1, 2, 3, 4, 5, 9, 11\}$.

Answer. 10, 55, 27

The BRICS Economies

BRICS is a partnership of economies united in 2010 with the purpose of highlighting and promoting investment opportunities. BRICS is considered as a competitor to the G7 alliance of leading advanced economies. All BRICS countries are members of G20, four of them are among the world's ten largest countries by population, area, and GDP, and three of them are superpowers. They take 27% of land and 42% of the global population. New countries are going to join BRICS in 2024. The acronym *BRICS* was coined in 2001 by Goldman Sachs economist Jim O'Neill who predicted fast-growing economies that will dominate by the middle of the 21st century.

The established New Development Bank or the NDB, is one of the BRICS initiatives. Its headquarters are in Shanghai, China. Its first regional office is in Johannesburg, South Africa. The NDB membership is open to all countries of the United Nations.

Problem. Using hints below, find how many countries are original BRICS members, will join BRICS in 2024, and are NDB members. Name these countries.

1. Space S consists of subsets B (BRICS members), A (joining members), and N (NDB members).
2. Elements included in a subset are added to get the requested numbers.
3. The sum of elements in $A \cup B \cup N$ is 13.
4. The sum of elements in $N \cap B$ is 5.
5. The sum of elements in $\bar{N} = N^C$ is 4.
6. The sum of elements in $(N / A) \cup B$ is 7.
7. $B \subset N$and $B \cap A = \varnothing$.

Answer. 5, 6, 9

Enclaves

Sovereign states surrounded by a single country are called *enclaves*. One of them is the Kingdom of Lesotho, located at the highest mountains in Southern Africa, the mountain of Maloti Mountains. The kingdom is surrounded (i.e., enclaved) by South Africa. It looks on the map like Lesotho has been inserted into the territory of South Africa.

After declaring independence from the United Kingdom in 1966, Lesotho became a fully sovereign state and is a member of the United Nations, the Commonwealth of Nations, the African Union, and the Southern African Development Community. Lesotho is a parliamentary or constitutional monarchy. The prime minister is the head of government, and the king of Lesotho is the head of the state with mostly ceremonial duties.

Problem. The number of enclaves in the modern word is the denominator of an irreducible fraction that represents the probability of selecting without replacement two red balls from a box that contains one white ball and five red balls. The number of countries that have enclaves is the numerator of this irreducible fraction. Find the number of enclaves and the number of countries that have enclaves. Can you name them?

Answer. 3, 2

Semi-enclaves

Semi-enclaves are sovereign states surrounded by one country and a coastline adjacent to international waters. For instance, the Southeastern Asian country Brunei is a semi-enclave because a Malaysian state encloses it except for a100-mile coastline along the South China Sea. Brunei is the only nation in the world that lies solely on an island, the island of Borneo.

Problem. The number of semi-enclaves coincides with the number of red balls in a box that has a total of 16 red, green, and blue balls. The box holds more than two balls of each color. The probability of selecting without replacement two balls, one is green and one is blue, is 1/6. Find the number of enclaves. Can you name them?

Answer. 3

Icelandic Glaciers

Iceland is a northern island country between the North Atlantic and Arctic Oceans, halfway between North America and Europe. The country is rich with everything, majestic mountain ridges, volcanos, geysers, glaciers, lakes, and waterfalls. The largest glacier of Iceland is Vatnajökull in the Vatnajökull National Park, one of the largest national parks in Europe. Active volcanoes, gorgeous canyons, largest glacier lagoons Jökulsárlón and Fjallsárlón, highest peak Hvannadalshnúkur, mighty waterfall Dettifoss, famous horseshoe-curved cliffs of Ásbyrgi, and vegetated oasis of Skaftafell can also be found in the park. Vatnajökull National Park was established in 2008 and became a UNESCO World Heritage Site in 2019.

Problem. Iceland consists of the main island of 101,826 km^2 and many minor islands. The minor islands altogether constitute 0.923% of the Iceland land. Lakes and glaciers cover 14.3% of the total Iceland surface. The total area of lakes is 2757 km^2. The glacier Vatnajökull with an area of 8100 km^2 occupies 58.05% of the Vatnajökull National Park territory. What percentage of the main Iceland island is covered by glaciers? What is the probability that a tourist in Iceland is inside Vatnajökull National Park?

Answer. 11.73%, 13.7%

Iceland Tourism

International tourism is a fast-growing industry in Iceland. The number of tourists arriving at the major Keflavik International Airport in Iceland increased from 700 thousand in 2012 to about 2 million in 2022.

Problem. 60% of all visitors to Iceland in 2022 came from six countries: the UK, Germany, France, Poland, the USA, and Canada. Some 5% more visitors come from Europe (the UK, Germany, France, and Poland) than from North America (the USA and Canada). The number of visitors from UK was 1.75

times more than from Germany. The number of visitors from Germany was 1.6 times more than from Poland. There were 1.1 times more visitors from France than from Poland. The number of Canadians was ten times less than Americans. Determine the two countries with the largest groups of visitors to Iceland. What is the probability that a randomly selected visitor is from these two countries?

Answer. 0.25, 0.14

Human Development vs. GDP

The United Nations' Human Development Index (HDI) measures a country's achievement in three basic dimensions of human development, long and healthy life, knowledge, and a decent standard of living. In 2015, the top ten countries with the best HDI among all 188 UN countries were Norway, Australia, Switzerland, Germany, Denmark, Singapore, Netherlands, Ireland, Iceland, and the USA. At the same time, the ten richest countries with the highest Gross Income (GDP) per capita included only three countries from the previous *Top Ten* list. The other seven were five Asian countries, Liechtenstein, and Luxembourg.

Problem. Using the minimum number of hints below, determine which three countries appeared in both Top Ten countries lists (with the best HDI and GDP). What are the probabilities that a randomly selected country belongs to one of the Top Ten lists and to both? What is the probability that a randomly selected country from Top Ten GDP list is in Asia?

A. Qatar, Brunei, Kuwait, United Arab Emirates, and Hong Kong were among the top ten countries with highest GDP in 2015.
B. Norway's GDP was sixth in the world in 2015.
C. The list of ten top-ranked HDI countries spread over four continents.
D. The list of ten top-ranked GDP countries spread over two continents.
E. Two of four European countries among ten top-ranked GDP countries are neutral.

Answer. 5.3%, 1.6%, 50%

The H-Day

Despite a rigorous several-year preparation, the H-day in Sweden was somewhat hectic but successful. Indeed, switching from driving on the left-hand side to the right of the road is not an easy task. The H-day, Dagen H, or

Högertrafikomläggningen, translated as the *right-hand traffic reorganization,* came, in spite of 83% of the Swedish population voting against the change. Considering international travelling and right driving in Norway and Finland, with which Sweden shares land borders, the Swedish Parliament initiated the change and associated educational campaign that included practical and phycological lessons, changing street sign, modifying vehicles, composing songs, manufacturing products displaying the Dagen H logo, and so on.

Problem. Find the year when Sweden switched from driving on the left side to the right side if two numbers formed from the first two digits and the last two digits of the year in the order they appeared in the year satisfy the following conditions:

A. The mean of the two numbers obtained from the first two digits and the last two digits of the year is 43.
B. The two numbers are odd.
C. The standard deviation of the set that consists of the two numbers is 24.
D. The variance of the set that contains the two numbers obtained from the first two digits and the last two digits of the year is 576.
E. The year when Sweden switched from driving on the left side of the road to driving on the right side is a composite number.
F. Driving in Sweden was on the left side of the road before September 3.

Choose the minimum number of hints.

Answer. 1967, A and C or D

Solutions

The United Kingdom. Figure 1.1 shows the diagram. The dash line outlines Ireland that contains both the Republic of Ireland and Northern Ireland. Great Britain is shown with the solid line. The United Kingdom of Great Britain and Northern Ireland (UK) is outlined with a dotted line.

The United Nations. Since elements 56, 68, and 72 are in $A \cup B$ (Hint A), and in $A \cap \overline{B}$ (Hint B), they are in the set A but not B (see Figure 1.2). Because 45, 69, and 79 are in $A \cap B$ (Hint C), these elements are in both sets A and B, and, in combination with the above, the set $A = \{45, 56, 68, 69, 72, 79\}$ and the set $B = \{5, 6, 45, 69, 79\}$. Finally, as 5 is not in $B \cap \overline{C}$, then 5 is the only element of the set C (see Figure 1.2). The sum of all elements in the set $C = \{5\}$ is 5, which gives the number of permanent members of the Security Council.

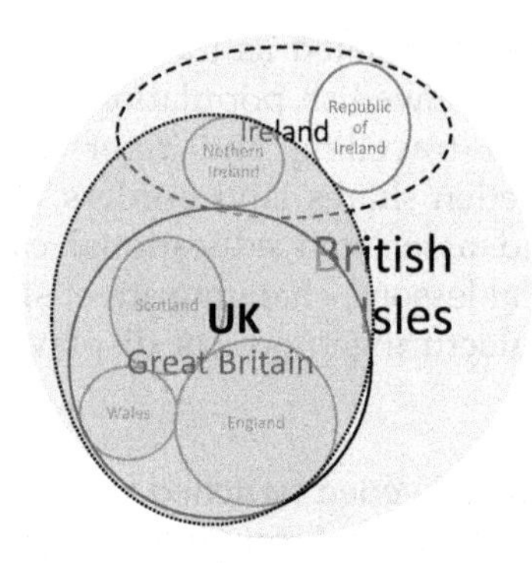

FIGURE 1.1

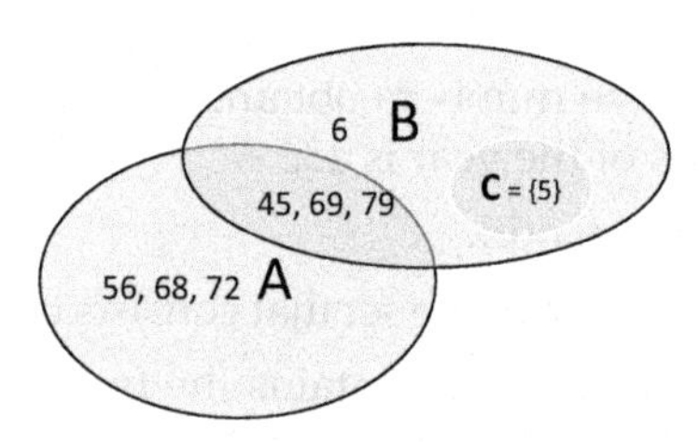

FIGURE 1.2

The product of the sum of all elements in the set $A = \{45, 56, 68, 69, 72, 79\}$and the sum of all elements in the set $C = \{5\}$is $(45 + 56 + 68 + 69 + 72 + 79)\cdot 5 = 1945$. $\overline{A} \cap B \cap \overline{C} = \{6\}$, then 6 is the only element in the set B that is neither in the set A nor in the set C.
The UN was established in 1945, upon ratification of the Charter by five permanent members of the Security Council, the US, the UK, France, the Soviet Union, and the Republic of China. The UN has six official languages, Arabic, Chinese, English, French, Russian, and Spanish.

Continental Unions. To solve the problem, let us do the following steps:

- $\overline{A} = \{2, 5, 11\}$, then $5\cdot 11 = 55$;
- $\overline{B} = \{2, 5, 9\}$, then $(2 + 9)\cdot 5 = 55$;
- $\overline{A} \cap \overline{B} = \{2, 5\}$, then $2\cdot 5 = 10$;
- $\overline{A \cap B} = \{2, 5, 9, 11\}$, then $2 + 5 + 9 + 11 = 27$.

The Association of Southeast Asian Nations unites 10 states. The African Union consist of 55 members.

The European Union is a union of 27 countries.

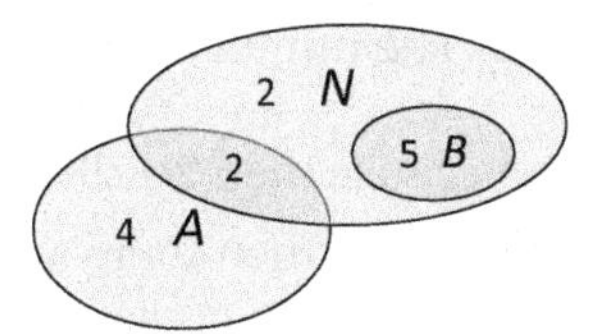

FIGURE 1.3

The BRICS Economies. The problem can be solved in different ways (see Figure 1.3). Only one of them is presented below. From Hints 3 and 5 follow that the sum of elements in N is 9. Hints 4 and 7 lead to 5 as the sum of elements in B. From Hints 5 and 6 follow that the sum of elements in A is $2 + 4 = 6$. Brazil, Russia, India, China, and South Africa are five members of BRICS. The acronym *BRICS* contains the first letters of these countries. Argentina, Egypt, Ethiopia, Iran, Saudi Arabia, and the United Arab Emirates are six countries that will join BRICS in 2024. Bangladesh, Egypt, the United Arab Emirates, and Uruguay are four countries that, in addition to five BRICS members, are nine members of NDB.

Enclaves. The probability of selecting two red balls from a box that contains one white ball and five red balls is

$$P = \frac{5}{6} \cdot \frac{4}{5} = \frac{2}{3},$$

where the probability of selecting the first red ball out of 6 is $5/6$. If the red ball has been selected, then only 4 red balls out of 5 are left and the probability of selecting the second red ball is $4/5$.

The Vatican City and Republic of San Marino are enclaved by Italy. The Kingdom of Lesotho is enclaved by South Africa.

Semi-enclaves. Let g and b be the number of green and blue balls. The probability of selecting two balls, green and blue, out of 16 balls in the box is

$$P = \frac{g}{16} \cdot \frac{b}{15} = \frac{1}{6} \text{ or } P = \frac{b}{16} \cdot \frac{g}{15} = \frac{1}{6}.$$

The order of selection of green and blue balls does not matter because the number of red balls is of our interest. Let us rewrite the last fraction as

$$P = \frac{bg}{16 \cdot 15} = \frac{bg}{2 \cdot 3 \cdot 2 \cdot 2 \cdot 2 \cdot 5} = \frac{1}{6} = \frac{2 \cdot 2 \cdot 2 \cdot 5}{2 \cdot 3 \cdot 2 \cdot 2 \cdot 2 \cdot 5} \Rightarrow bg = 2 \cdot 2 \cdot 2 \cdot 5,$$

and consider different options. The numbers of green and blue balls can be

- (2, 20) – does not work, because then $2 + 20 > 16$;
- (4, 10) – does not work, because then there are only $16 - (4 + 10) = 2$ red balls, that contradicts the condition of *more than two balls of each color*;
- (8, 5) – works and leads to $16 - 8 - 5 = 3$ red balls.

Monaco in Europe, Gambia in Africa, and Brunei in Asia are the only semi-enclaves in the world.

Icelandic Glaciers. The solution includes the following steps:

- The area 101,826 km^2 of the main island takes (1 – 0.00923)%.
- The area of the country is 101,826 km^2 /(1 – 0.00923) = 102775 km^2.
- The total percentage of lakes is 2757/102,775 = 2.68%.
- The total percentage of glaciers is 14.3% – 2.68% = 11.62%.
- The total percentage of glaciers in the main island is 11.62/(1 – 0.00923) = 11.73%.
- The area of the Vatnajökull National Park is 8100/0.5805 = 13,952 km^2.
- The percentage of the Vatnajökull Area in the main island is 13,952 km^2 /101,826 km^2 = 13.7%.
- The probability that a tourist is in the Vatnajökull National Park is 0.137.

Therefore, the glaciers cover 11.73% of the main island. The Vatnajökull NP covers 13.7% of the island.

Iceland Tourism. Let x and y be the percentage of visitors from Poland and USA. Then,

$$x(1.75 \cdot 1.6 + 1.6 + 1.1 + 1) = 6.5x$$

is the percentage of European visitors and $1.1y$ is the percentage from USA and Canada. Hence, we get two equations

$$6.5x + 1.1y = 60 \text{ and } 6.5x - 1.1y = 5 \Rightarrow x = 5 \text{ and } y = 25.$$

Then, $2.8x = 14\%$ and $1.1y = 27.5\%$.

Therefore, 25% of Iceland visitors came from USA and 14 % came from UK. The probability that a randomly selected visitor is from the USA is 0.25 and 0.14 from UK.

Human Development vs. GDP. By Hint D, Australia and USA were not among ten top-ranked GDP countries in 2015. By Hint C, Singapore belongs to both *Top Ten* lists, which means that two other countries in both lists are European. By Hint B, one of them is Norway. By Hint E, two of the listed European countries must be neutral, which are Switzerland and Liechtenstein. Therefore, Switzerland is the third country in both lists.

The probability that a UN country is in at least one of the *Top Ten* list is 10/188 = 0.053.

The probability that a UN country is in both *Top Ten* lists is 3/188 = 0.016.

If a random country from Top Ten GDP list is picked up, then it is in Asia with the probability of 5/10 = 0.5.

The H-Day. Let x and y be two numbers obtained from the first and last two digits of the year, $x < y$. From the formula of their mean

$$m = \frac{x+y}{2} = 43, \text{ we get } x = 2 \cdot 43 - y$$

that substituted to the formula for standard deviation gives

$$s = \sqrt{\frac{(m-x)^2 + (m-y)^2}{2}} = \sqrt{\frac{(m-y)^2 + (m-y)^2}{2}} = |m-y| = y - m = y - 43 = 24.$$

To go from $|m-y|$ to $(y-m)$, we use $x < m < y$. Then $y = 67$ and $x = 19$.

Note. The formula for a variance instead of the one for a standard deviation can be used.

The driving on the left side of the road to driving on the right side was switched in Sweden on September 3, 1967.

Sources

en.wikipedia.org/wiki/BRICS

en.wikipedia.org/wiki/New_Development_Bank

Statistics Iceland, 2023

en.wikipedia.org/wiki/Enclave_and_exclave

en.wikipedia.org/wiki/United_Nations

borgenproject.org/10-cool-facts-about-the-united-nations/

en.wikipedia.org/wiki/African_Union

en.wikipedia.org/wiki/European_Union

en.wikipedia.org/wiki/ASEAN

en.wikipedia.org/wiki/United_Kingdom

en.wikipedia.org/wiki/Enclave_and_exclave

en.wikipedia.org/wiki/Lesotho

The Icelandic Tourist Board, 2023

www.autoweek.com/car-life/a1829546/50-years-ago-today-sweden-switched-driving-right-driving-left

en.wikipedia.org/wiki/Dagen_H

1.6 The Most Extreme

The Smallest Country in Oceania

The Republic of Nauru with just 11,000 citizens is the smallest country in Oceania, the smallest island nation, and the smallest republic in the world. Although Nauru is the least populous country in the Pacific Ocean, and the third least populous country in the world, the country is among the most densely populated countries with over 600 people per km^2. Due to its amazing coastline, Nauru is also called the *Pleasant Island*.

Problem. Find the area c km^2 of Nauru, the length b km of its coastline, and the place a Nauru takes among the world's most densely populous countries, if a, b, and c are coefficients of the equation of the parabola

$$y = ax^2 - bx + c,$$

that has the minimum at (1.5, –2.5) and passes through (3, 20).

Answer. 20 km^2, 30 km, 10

The Longest Border

Canada follows Russia in the list of countries with the largest areas, but it has the longest coastline and the longest international border with one country, the United States. Canada is washed by the Pacific Ocean in the west, the Arctic Ocean in the north, and the Atlantic Ocean in the east. The length of Canadian coastline is greater than the following longest coastlines of Indonesia, Norway, and Russia taken together.

Problem. How long are the Canadian coastline and its border with the United States if the border length in thousand kilometers is p and the Canadian coastline stretches pq thousand kilometers. The values p and q are coefficients of the function

$$y = x^3 + px^2 + rx + q,$$

that reaches its maximum at (–5, 52), minimum at (–1, 20), and has the y-intercept at (0, 27). Sketch the graph of the function.

Answer. 9000 km, 243,000 km

The Deepest River

The Congo River or the Zaire River was named after the ancient Kingdom of Kongo (now territory of Angola) founded by the *KiKongo* speaking people. The Portuguese at the middle of the 19th century spelled *Kongo* as *Congo*. After passing Democratic Republic of Congo, Republic of Congo, Cameroon, the Central African Republic, Burundi, Tanzania, Zambia, and Angola, the Congo River brings its water to the Atlantic Ocean. It is the deepest river, has the second largest drainage basin after the Amazon River, and is among the longest rivers in the world. This winding river is the only major river that crosses the Equator more than once.

Problem. How deep is the Congo River if its maximal depth in meters can be presented as the absolute value of the product of the coordinates of the maximum point of the function

$$y = -x^2 + px + q,$$

that passes through (0, –122) and (15, –47)? How many times does the river cross the Equator if this number is the lowest value that x takes when $y = -86$?

Answer. 220 meters, 2

The Largest River

The Amazon River flows through Brazil, Peru, Bolivia, Colombia, Ecuador, Venezuela, and Guyana bringing its water to the Atlantic Ocean. With its main headstreams Ucayali, Tambo, Ene, and Mantaro, the Amazon River is the world's largest river by discharge volume of water and drainage basin, which is approximately the size of Australia. There is also a claim that the Amazon River is the world's longest, not the second longest river after the Nile. The *River Marañón* in South America was named *Rio Amazonas* by the Spanish and Portuguese after furious attacks by native warriors on the Francisco de Orellana's expedition in the 16th century. Native women or males with long hair resembled the *Amazon warriors* from the Greek mythology. Another version states that the name is derived from the Native American word *amassona*, meaning *boat destroyer*.

Surprisingly, there is no bridge across the Amazon River.

Problem. The length of the Amazon River in kilometers is the same as the numeral value of the maximum area that a rectangle with the perimeter of 320 units can have. How long is the Amazon River?

Answer. 6400 km

The Tallest Mountain

Mount Everest in Nepal and Tibet is the highest mountain above sea level. This astonishing mountain in the Himalayas is sacred to many nations. The Tibetans call it *Holy Mother*, Nepalis prefer *Goddess of the Sky* or *the Head in the Great Blue Sky*, the Chinese refer to it as *Holy Mother Peak* or *Holy Mountain*. At the middle of 19th century, the peak was named after British surveyor Sir George Everest even with his objection. Despite dangerous climbs, extreme cold, fierce winds, limited oxygen, and many deaths, this spectacular mountain attracts many climbers each year.

This stunning natural marvel inspires many legends and myths. It is also said that Yeti, Ghosts of Climbers, Bigfoot, and other mythical monsters have been seen there. A story of Mont Everest appears in *Peaks: Nothing Is Impossible*, *Beyond the Edge*, *The Conquest of Everest*, *The Epic of Everest*, *Everest*, *The Himalayas*, *Sherpa*, *The Wildest Dream*, *Yeti Obhijaan*, and many other documentaries and movies.

Problem. The height of Mount Everest above sea level in meters can be presented by a four-digit number. Its third (tens) and forth (units) digits are x and y coordinates of the maximum point of the function

$$y = -x^4 + 4x^3 + 12x^2 - 32x - 72$$

located in the first quadrant. The second (hundreds) digit is the y coordinate of the maximum point located in the second quadrant. The first (thousands) digit is the maximum among the units, tens, and hundreds of the number. How tall is Mount Everest?

Answer: 8848 m

The Wettest Capital

The climate on our planet varies dramatically from cold to heat and from dry to wet. It is a mix of different temperature and precipitation. Mawsynram in Meghalaya, India, is the wettest place on the Earth, with an average annual rainfall of 11,871 mm. With tropical rainforest climate and only two rainy seasons, Tutunendo in Colombia has an average annual rainfall of 11,770 mm. Liberia's capital, Monrovia, with an annual average rainfall of over 4500 mm competes with Guinea's capital, Conakry City, for the title of the wettest capital on the planet.

Problem. The integer part of the average annual rainfall amount in Conakry City in meters is the absolute value of the x-coordinate of the point where the minimum of the function

$$y = 3x^2 + px + q,$$

that passes (0, 20) and the maximum of the function

$$y = -2x^2 + rx + s,$$

that crosses (0, –25), meet. The decimal of this rainfall amount is the absolute value of the y-coordinates of this point of intersection. What is the average annual rainfall amount in Conakry City?

Answer. 3.7 meters

Basalt Columns

The appearance of basalt columns is caused by centuries of volcanic eruptions during which the lava hardened, cooled, and then cracked into vertical cliffs, columnar tiles, or terraced steps that often descend into the water. They can be found around the world. Thousands of visitors admire breathtaking views and legends about the Giant's Causeway in Northern Ireland, the Devils Postpile in the USA, the Basaltic Prisms of Santa María Regla in Mexico, Fingal's Cave in Scotland, Svartifoss in Iceland, Cape Stolbckatiy cliffs in Russia, and Organ Pipes in Namibia, just to name a few. These impressive natural marvels have inspired composers, poets, and artists to create memorable pieces of work.

The gorgeous geologic wonders have a lot in common. In particular, their cross sections are most often regular polygons connecting to each other.

Problem. Cross sections of basalt columns are of regular polygons. Nature does not work in vain. The polygons are closely connected leaving no space between them. Describe a regular polygon in the cross section of a basalt column that has the maximum area under the perimeter P. Find examples of other natural phenomena that have the same geometric shape.

Answer. Hexagon

The Longest Tunnel

Highways in mountains cannot be imagined without well-developed tunnels that significantly shorten the travel time. These remarkable engineering constructions connect cities and countries, go through the mountains and beneath the water. The 50-km underwater railway Channel Tunnel connects the shores of the UK and France. The European longest roadway and railway between Denmark and Sweden would not be possible without the 4-km Drogden tunnel. The 2.5-mi Whittier tunnel that goes through an Alaskan mountain is the longest combined rail/highway tunnel in North America.

The longest road tunnel on the planet, the Laerdal tunnel in Norway, was open in 2000. It is a very high-tech tunnel with daylight illumination, caves to reverse direction, large caverns to stop and relax, and internet connection. Ventilation and purification maintain high air quality there. Although there are no emergency exits, emergency phones are placed 125 m apart.

Problem. Find the length of the Laerdal tunnel in meters if it can be presented as *uetln* where the letters e, l, n, t, and u, stand for digits 1, 2, 3, 4, 5 in the way that a value of the fraction f

$$f = \cfrac{1}{t + \cfrac{1}{u + \cfrac{1}{n + \cfrac{1}{e + \cfrac{1}{l}}}}}$$

takes its maximal value. Letters are given in alphabetical order.

Answer. 24,513 m

The Shortest Land Border

Which country has the longest land border overall? If you think that it is Russia, which is the world's largest country by area, then think again. Canada, the second largest country by area, shares the longest international 8893-km land border with the United States. It consists of two sections, Canada's border with the contiguous United States to its south, and with the US state of Alaska to its west.

Is the US the only country that Canada has the land border with? If you think *yes*, then think again. Canada has a land border with a European country, and it is… the Kingdom of Denmark. Yes, the world's longest sea border between Canada and Denmark passes through a tiny uninhabited rock Hans Island in the Arctic. Remarkably, the only other land border that Denmark has is with Germany.

Problem. How long is the Canadian–Denmark land border if its length in miles is the solution x to the equation

$$\min(3^{125x}, 4^{100x}) = \max(3^{100}, 4^{75})?$$

Answer. 0.8 mi

The Ring of Fire

About 75% of the Earth's volcanic eruptions and 90% of earthquakes are along the *Ring of Fire*, also referred to as the *Pacific Ring of Fire*, the *Rim of Fire*, the *Girdle of Fire*, or the *Circum-Pacific belt*. Such seismic activities are caused by the location of this horseshoe-shaped 25,000 mi by 310 mi belt at the boundaries of Pacific, Juan de Fuca, Cocos, Indian-Australian, Nazca, North American, and Philippine Tectonic Plates that constantly move affecting the Pacific coasts of South America, North America, Asia, and islands in the Pacific Ocean. The world's deepest ocean trench, Marianna Trench, is also in the Ring of Fire.

Ancient Greeks and Romans considered volcanoes as fires burning within the Earth. This belief coined the name of the *Ring of Fire* in a book published in 1906.

Problem. The number of volcanoes along the Ring of Fire is a three-digit number with digits represented as the bases of the powers 2^{36}, 4^{13}, 5^{12}, after the powers have been arranged in the increasing order. The number of countries in the Ring of Fire is the product of the middle digit and the average of the outer two digits of the number of volcanos. How many volcanos are along the Ring of Fire? What volcanoes do you know? How many countries are located in this seismic zone? Name the countries that you know.

Answer. 452, 15

Solutions

The Smallest Country in Oceania. The vertex of a parabola $y = ax^2 - bx + c$, which is also its minimum point, is determined by

$$x = \frac{b}{2a} \text{ and } y = \frac{4ac - b^2}{4a}.$$

Then

$$1.5 = \frac{b}{2a} \text{ or } b = 3a \text{ and } -2.5 = \frac{4ac - b^2}{4a} = \frac{4ac - 9a^2}{4a} \text{ or } 4c = 9a - 10.$$

Substituting the expressions of b and c to the equation of a parabola and using the point (3, 20), we obtain

$$20 = a \cdot 3^2 - 3a \cdot 3 + (9a - 10)/4 \text{ or } a = 10.$$

Then, $y = 10x^2 - 30x + 20$.

The smallest country in Oceania with the territory of just 20 km² (8.1 sq mi) with 30 km coastline, Nauru, is the 10th most densely populated country in the world.

The Longest Border. Since the y-intercept is at (0, 27), then $q = 27$ (see Figure 1.4). Indeed, substituting $x = 0$ to $y = x^3 + px^2 + rx + q$ gives $y = 0^3 + p \cdot 0^2 + r \cdot 0 + q = 27$. Applying (–5,52) and (–1, 20) to $y = x^3 + px^2 + rx + 27$, we get the system of two linear equations with respect to p and r

$$\begin{cases} (-1)^3 + p \cdot (-1)^2 + r(-1) + 27 = 20 \\ (-5)^3 + p \cdot (-5)^2 + r(-5) + 27 = 52 \end{cases} \text{ or } \begin{cases} p - r = -6 \\ 25p - 5r = 150 \end{cases},$$

that has the solution $p = 9$, $r = 15$.

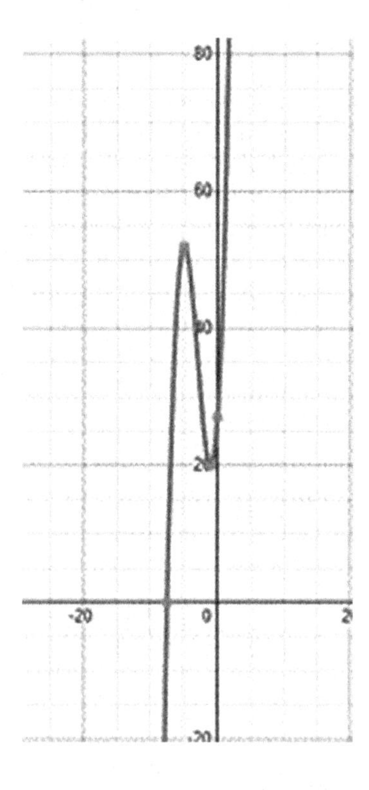

FIGURE 1.4

The Canadian border with the United States is 9000 km, and the Canadian coastline is 243,000 km.

The Deepest River. Having the point (0, –122), it is obvious that $q = -122$. Substituting the second point (15, –47) to the equation, we can find p:

$$-47 = -15^2 + p \cdot 15 - 122 \Rightarrow p = 20.$$

The maximum point of the parabola

$$y = -x^2 + 20x - 122 = -(x - 10)^2 - 22$$

that opens downward is (10, –22). Solving the quadratic equation

$$-86 = -x^2 + 20x - 122 \Rightarrow x = 10 \pm \sqrt{100 - 36}, \text{ we get } x = 2, x = 18.$$

The lowest value is $x = 2$.

The Congo River is 220 m deep and crosses the equator twice. From the southern hemisphere, the Congo River takes a northeastward route to the Northern Hemisphere and then starts flowing towards southwest back into the Southern Hemisphere.

The Largest River. Let l and w be the length and the width of a rectangle. Then, $2l + 2w = 320$. The area of the rectangle is

$$A = lw = l \cdot (160 - l) = -l^2 + 160l = -(l^2 + 2 \cdot 80l + 80^2) + 80^2 = -(l - 80)^2 + 6400.$$

The graph of the quadratic equation for the area is a parabola that opens downward with the maximum point at its vertex (80, 6400). The rectangle with the maximum area and given perimeter is a square with an area of 6400 sq. units.

The Amazon River is 6400 km or 3976 miles long.

The Tallest Mountain. The fourth-degree parabola $y = -x^4 + 4x^3 + 12x^2 - 32x - 72$ opens downward. Let us rewrite the function so that its leading coefficient is positive and rearrange terms:

$$\begin{aligned} y &= -(x^4 - 4x^3 - 12x^2 + 32x + 72) \\ &= -(x^4 - 4x^3 + 6x^2 - 4x + 1 - 18x^2 + 36x - 18 + 89) \\ &= -((x^4 - 4x^3 + 6x^2 - 4x + 1) - 18(x^2 - 2x + 1) + 89) \\ &= -((x - 1)^4 - 18(x - 1)^2 + 89). \end{aligned}$$

Substitution of $t = (x - 1)^2$ reduces the above polynomial function of the fourth degree to the quadratic function

$$y = -(t^2 - 18t + 89) = -(t^2 - 2{\cdot}9t + 81 + 8) = -(t - 9)^2 + 8.$$

The last function is represented as a parabola that opens down and has a maximum at (9, 8).

The back substitution $x = \pm\sqrt{t} + 1$ gives two maximum points $x = -2$ and $x = 4$. Thus, the original polynomial function of the fourth degree has two maximum points (–2, 8) and (4, 8). Then, following the problem statement, the third and fourth digits are 4 and 8, the third digit is 8, and the maximum of these digits is 8. Hence, the number is 8848.

The Mount Everest's peak is 8848 meters (29,029 feet) above sea level.

The Wettest Capital. Both functions are represented by parabolas. One of them is open up, another is open down. Let us rewrite the equations using the point (a, b) of their intersection, that is the maximum point of one function and the minimum point of another,

$$y = 3(x - a)^2 + b \text{ and } y = -2(x - a)^2 + b.$$

Then the given points (0, 20) and (0, –25) lead to the system

$$\begin{cases} 20 = 3a^2 + b \\ -25 = -2a^2 + b' \end{cases}$$

that has the solution $a = \pm 3$ and $b = -7$. Taking the absolute values, the number is 3.7

Conakry City experiences an average of 3700 mm of rain annually.

Basalt Columns. Let an integer $n, n \geq 3$, be the number of sides of a regular polygon and a be its side. The condition $n \geq 3$reflects a triangle as the polygon with the smallest number of sides. The interior angle of a regular polygon is $\angle B = \frac{n-2}{n} \cdot 180^0$.

Remembering that polygons are closely connected leaving no space between them, the number $f(n)$ of identical regular polygons connected to each other can be found as

$$f(n) = \frac{360^\circ}{\angle B} = \frac{n \cdot 360^\circ}{(n-2) \cdot 180^\circ} = \frac{2n}{(n-2)}.$$

The number $f(n)$ is integer only when $n = 3, 4,$ or 6.

The perimeter and area of a regular polygon are

$$P = na \text{ and } A = \frac{na^2}{4}\cdot\cot\frac{\pi}{n} \Rightarrow A(n) = \frac{P^2}{4n}\cdot\cot\frac{\pi}{n}.$$

Let us create a table with all these values.

n	$f(n)$	$\angle B$	$A(n)$
3	6	$60^\circ \frac{\pi}{3}$	$\frac{P^2\sqrt{3}}{36}$
4	4	$90^\circ \frac{\pi}{2}$	$\frac{P^2}{16}$
6	3	$120^\circ \frac{2\pi}{3}$	$\frac{P^2\sqrt{3}}{24}$

and compare the areas of polygons after using the common denominator of 144 and presenting the fractions as

$$\frac{P^2\sqrt{3}}{36} = \frac{4P^2\sqrt{3}}{114} < \frac{P^2}{16} = \frac{9P^2}{144} < \frac{P^2\sqrt{3}}{24} = \frac{6P^2\sqrt{3}}{144},$$

Then, we get

$$\frac{P^2\sqrt{3}}{36} < \frac{P^2}{16} < \frac{P^2\sqrt{3}}{24}.$$

Therefore, $n = 6$.

Cross sections of basalt columns worldwide are mostly hexagons. Hexagonal patterns are prevalent in nature due to their efficiency.

The Longest Tunnel. There are several ways to solve the problem. One of them is based on the property that a unit fraction is greater if its denominator is smaller. Thus, $t = 1$, $u = 2$, $n = 3$, $e = 4$, $l = 5$, and the number *uetln* is decoded as 24513.

The 24.5-km Laerdal tunnel in Norway is the longest tunnel on the planet.

Shortest Land Border. To find the value of $\max(3^{100}, 4^{75})$, let us present 3^{100} as $3^{4\cdot 25} = (3^4)^{25} = (81)^{25}$ and 4^{75} as $(4^3)^{25} = (64)^{25}$. Since $1 < 64 < 81$, we conclude that $\max(3^{100}, 4^{75}) = 3^{100}$.

Analogously, rewriting 3^{125x} as $3^{5\cdot 25x} = (3^{5x})^{25} = (243^x)^{25}$ and 4^{100x} as $(256^x)^{25}$ and considering that $1 < 243 < 256$, we get $\min(3^{125x}, 4^{100x}) = 3^{125x}$.

Hence $\min(3^{125x}, 4^{100x}) = \max(3^{100}, 4^{75})$ turns to $3^{125x} = 3^{100}$ that leads to $125x = 100$ or $x = 4/5 = 0.8$.

The 0.8-mi Canada–Denmark land border divides the Hans Island almost in halves.

The Ring of Fire. Let us compare 2^{36} and 5^{12} first. Using $a^{mn} = (a^m)^n$, we can write $2^{36} = (2^3)^{12} = 8^{12}$. Because $8 > 5 > 1$, then $8^{12} > 5^{12}$ and $2^{36} > 5^{12}$. Comparing $2^{36} = 4^{18}$ and 4^{13} we get $2^{36} > 4^{13}$. Thus, 2^{36} is the largest.

Remembering that $y = a^x$ is an increasing function if $a > 1$, presenting 4^{13} as $4^8 \cdot 4^5 = 65536 \cdot 4^5$ and 5^{12} as $5^7 \cdot 5^5 = 78125 \cdot 5^5$, and comparing $65536 \cdot 4^5$ and $78125 \cdot 5^5$, we conclude that $4^{13} < 5^{12}$. Thus, $4^{13} < 5^{12} < 2^{36}$ and the number is 452. The product of the middle digit and the average of the outer two digits $5 \cdot (4 + 2)/2 = 15$.

Mount St. Helens, Ojos del Salado, Llullaillaco are among 452 volcanoes located along the Ring of Fire.

Indonesia, New Zealand, Papua New Guinea, Philippines, Japan, United States, Chile, Canada, Guatemala, Russia, Peru, Solomon Islands, Mexico, and Antarctica are among 15 countries of the Ring of Fire.

Sources

www.nationsonline.org/oneworld/oceania.htm; en.wikipedia.org/wiki/Nauru

worldpopulationreview.com/country-rankings/countries-by-coastline

www150.statcan.gc.ca/n1/pub/11-402-x/2011000/chap/geo/geo01-eng.htm

en.wikipedia.org/wiki/Congo_River

en.wikipedia.org/wiki/Amazon_River

en.wikipedia.org/wiki/List_of_rivers_by_length

timesofindia.indiatimes.com/travel/destinations/in-pictures-the-wettest-places-on-earth

facts.net/country-facts/

www.treehugger.com/strangest-basalt-column-landscapes-earth-4869700

education.nationalgeographic.org/resource/ring-fire/

en.wikipedia.org/wiki/Ring_of_Fire

science.howstuffworks.com/environmental/earth/geology/ring-of-fire

en.wikipedia.org/wiki/L%C3%A6rdal_Tunnel

www.worldatlas.com/articles/countries-with-the-longest-land-borders.html

1.7 UNESCO Sites with Logarithms

UNESCO World Heritage Sites

World Heritage Sites are various natural and man-made outstanding landmarks chosen by UNESCO for their cultural, historical, scientific, religious, and/ or natural importance. Remarkable cities, ancient structures, worship sites, national parks, and other places are carefully selected to be included in the World Heritage List established in 1978. They belong to all the people, regardless of their location.

Problem. How many remarkable landmarks were included in the first UNESCO World Heritage List, if this number can be presented as $\log_{\sqrt[3]{a}} b^{\log_{\sqrt{b}} a^2}$? List those that you know.

Answer. 12

Countries with the Most Heritage Sites

There are more than 1000 UNESCO World Heritage Sites worldwide. They come in all shapes and sizes. The Phoenix Islands Protected Area in Kiribati is the largest World Heritage Site. It encompasses over 408,250 km^2, while the smallest one with the area of just 0.0002 km^2 is the Holy Trinity Column in Czech Republic. Some UNESCO Heritage Sites are in one country, while others spread over several countries. The Struve Geodetic Arc comprises the most countries, namely, Belarus, Estonia, Finland, Latvia, Lithuania, Norway, Moldova, Russia, Sweden, and Ukraine. The most visited World Heritage Site is the Banks of the Seine in Paris, France, while Italy has the largest number of Heritage Sites.

Problem. Find the number of UNESCO World Heritage Sites in Italy, if this number is a positive integer solution to the equation

$$\ln(x^2 - 56x - 57) = \frac{\ln(59 - x)}{\ln(x^2 - 58)} + \ln 59.$$

What UNESCO World Heritage Sites in Italy have you visited?

Answer. 58

World Heritage Cities

Over 300 cities across the globe are thoroughly preserved as World Heritage Sites. One of the oldest Spanish towns, Cordoba, has the highest number of World Heritage Sites in one city. Built as a Roman settlement, then by the Umayyad-Caliphate, and later becoming Christian again, Cordoba proudly preserves its mosques, cathedrals, fortresses, castles, festivals, and other historic and cultural treasures.

Vatican City and Paphos are cities that are UNESCO World Heritage Sites by themselves. Paphos is a town in Cyprus, an island country in the eastern Mediterranean Sea that has Greek and Turkish as its official languages. Its capital, Nicosia, is the only capital that has been divided by *The Green Line* between two nations, the Republic of Cyprus and the Turkish Republic of Northern Cyprus. Antique Cyprus has maintained close relations with Greece since Alexander the Great seized the island in 333 BCE but has *never* been a Greek Island. It is believed that the world's oldest wine was made in Cyprus over 5000 years ago.

It is interesting that Cyprus and Kosovo are the only two nations that have their maps on their national flags.

Problem. Cypress has $\log_{\frac{1}{8}} \log_4 \sqrt{\sqrt[8]{\sqrt[16]{2}}}$ UNESCO World Heritage Sites, while Cordova in Spain has one more. How many UNESCO World Heritage Sites do Cyprus and Cordova have? List those that you know.

Answer. 3, 4

Countries with the Least Heritage Sites

The UNESCO World Heritage List consists of sights with significant natural or human-made values. Each country is proud of having its landmarks mentioned there. Only countries that sign the 1978 convention of the World Heritage Committee are eligible to submit their applications. Being among countries-signatories, Bhutan, Guinea-Bissau, Guyana, Liberia, Sierra Leone, Somalia, and South Sudan have also nominated sites. Guyana and Liberia nominated their archeological structures, ancient monasteries, wildlife sanctuaries, and national parks. Bhutan submitted eight sites to be considered for the World Heritage List.

Problem. Bhutan, Guinea-Bissau, Guyana, Liberia, Sierra Leone, Somalia, and South Sudan have the same number of UNESCO World Heritage Sites. Find this number presented as the product

$$\log 2^{10} \cdot \log 2^9 \cdot \log 2^8 \cdot \ldots \cdot \log 2^2 \cdot \log 2^1 \cdot \log 2^0 \cdot \log 2^{-1} \cdot \log 2^{-2} \cdot \ldots \cdot \log 2^{-9} \cdot \log 2^{-10}.$$

List them.

Answer. 0

Restricted UNESCO Heritage Sites

UNESCO World Heritage Sites preserve the legacy from the past and outstanding natural landscapes to pass them on to future generations. The United States has dedicated the largest area to be listed as World Heritage Sites, while Vatican City is the country with the highest percentage of over 100% of its area to be included to the UNESCO World Heritage List, just because some of its landmarks are counted twice, as Vatican City itself and as the Historic Center of Rome in Italy.

Although World Heritage Sites are open to all people, some of them are hardly visited, others are either forbidden for women or banned for all visitors. The island Surtsey formed by volcanic eruptions from 1963 to 1967 is just 32 km south of Iceland but it is open only for scientists several times in a decade.

Problem. How many UNESCO World Heritage Sites are forbidden for women if this number is the same as $\sqrt[3]{2^{\frac{\log_{81} 1}{\log_9 4}} \cdot 27^{\frac{1}{\log_4 9}}}$. List them.

Answer. 2

Losing Heritage Status

The government and people work extremely hard to prepare their landmarks for a nomination, though being mentioned in the UNESCO World Heritage List does not guarantee their permanency. A site can be completely or partially removed from the list if it is not properly managed or its conservation obligations are not fulfilled. The Arabian Oryx Sanctuary in the Central Desert and Coastal Hills of Oman was entirely removed from the World Heritage List, while only the cathedral was removed from the Bagrati Cathedral in Georgia leaving its monastery in the list.

Problem. How many countries have sites permanently removed from the UNESCO World Heritage List if this number is the same as $\log_2 1000 - \frac{1}{\log_5 \sqrt[3]{2}}$? List them.

Answer. 3

Disjointed UNESCO Heritage Sites

The exceptional examples of the world's cultural and natural heritage are carefully selected to be presented in the UNESCO World Heritage List. Currently the list includes over 1000 sites located in 167 countries. Most of them are registered as cultural properties, others are recorded as natural and mixed ones. The list constantly changes with the addition of new astonishing landmarks. The most number of sites were added in 2000, while 1989 was the year with the fewest new sites appeared on the list.

Heritage Sites are unique and diverse, varying from ancient ruins and modern cities to uninhabited deserts, unique mountains, and water resources. Wrangell-St. Ellias National Park in Alaska, USA, is one of the largest heritage areas. Spreading from Granada to Catalonia along the coast of Spain, the Rock Art of the Mediterranean Basin on the Iberian Peninsula is the most disjointed World Heritage Site.

Problem. The Rock Art of the Mediterranean Basin has many locations. This number of locations can be presented as a three-digit number. Its hundreds x and ones y satisfy the system

$$\begin{cases} \log_x y = \log_y x \\ 3x - 2y = 7 \end{cases}.$$

Its tens digit is twice x/y.

The numbers of sites added to the World Heritage list in 2000 and 1989 are $(9y - 2)$ and x correspondingly. Find the number of places of the Rock Art of the Mediterranean Basin and the numbers of landmarks added to the UNESCO World Heritage List in 2000 and 1989.

Answer. 727, 61, 7

Heritage City on a Coral Reef

Micronesia consists of four main archipelagos and numerous small volcanic and coral islands in the western Pacific Ocean north of Australia and east of Asia. People inhabited these beautiful islands a long time ago, while a Spanish expedition led by Ferdinand Magellan discovered them in 1521. Micronesia was later colonized by Spanish, German, British, Americans, and other countries and empires. Currently, there are several independent island nations in Micronesia. The least visited of them, Federated States of Micronesia, which should not be confused with the wider geographic region of Micronesia, is home to the only ancient city ever built on a coral reef. This city of Nan Madol has been recognized as a UNESCO World Heritage Site.

Problem. The sum of $\log_3 40.5$ and $\log_3 54$ provides the number of countries in Micronesia, while their difference is $\log_3 4$ more than the number of UNESCO World Heritage Sites in Federated States of Micronesia. How many countries are in Micronesia? How many UNESCO World Heritage Sites are in Federated States of Micronesia? List those that you know.

Answer. 7, 1

UNESCO Heritage Lakes

The Republic of Malawi is a small landlocked country in Southeastern Africa. It used to be called *Nyasaland* under the British rule. The country takes its name, meaning *fire flames* either from the reflection of the rising sun on the waters of the picturesque Lake Malawi or from the powerful Maravi Kingdom that occupied the region in the 16th century. Despite being one of the poorest African countries, Malawi is often called *the Warm Heart of Africa* due to the incredible hospitality of its smiling citizens.

Malawi is a member of the United Nations, the African Union, and other international organizations. The country has two UNESCO World Heritage Sites, Lake Malawi National Park and the 127 site-Chongoni Rock Art Area with the richest collection of rock art in Central Africa. Lake Malawi, also known as *Lake Nyasa* in Tanzania and *Lago Niassa* in Mozambique, is the African Great Lake and is among the world's largest freshwater lakes by volume and area. The lake has more fish species than any other lake in the world. The UNESCO Lake Malawi National Park is entirely in Malawi.

It is interesting that Lake Malawi is a meromictic lake because its water layers do not mix.

Problem. With an area of 11500 sq. miles, Lake Malawi is approximately $\log_{125} 5 \cdot \log_{81} 27$ of Malawi's area. What is the area of Malawi?

Answer. 46000 sq. mi

UNESCO Archipelagos

An archipelago is a group of islands closely scattered in the ocean. The largest group of islands on the planet is Malay Archipelago. It contains more than 17,000 islands of Indonesia and more than 7,000 islands of the Philippines. Some countries, like Sweden and Canada, have one or more archipelagos, while archipelagic states, like Fiji and Marshall Islands, are entirely on one or several archipelagos.

The unique landscape and biodiversity of marine life with rich and diverse flora and fauna do not appear anywhere else in the world. Therefore, some archipelagos are listed as UNESCO World Heritage Sites. Among them are 19 islands of the Galápagos Archipelago of Ecuador in the Pacific Ocean at the confluence of three ocean currents, 250 km long Socotra Archipelago of Yemen in the Indian Ocean, 5600 islands of the Kvarken Archipelago of Finland, a cluster of dozens of islands of the Vega Archipelago of Norway just south of the Arctic Circle, and four remote islands of the Archipiélago de Revillagigedo of Mexico in the eastern Pacific Ocean.

Problem. The number of archipelagos in the Republic of Seychelles is the maximum solution to the logarithmic equation

$$\log_2 \frac{x}{8} = \log_x \frac{1}{4},$$

while the value of its smallest solution provides the number of archipelagos of the Republic of the Marshall Islands, the Republic of Mauritius, the Republic of Cabo Verde, the Republic of Vanuatu. The Republic of Fiji consists of one less archipelago than Mauritius, while Solomon Islands is located on one more archipelago than Seychelles. Find the number of archipelagos that belong to each archipelagic country listed above.

Answer. 1, 2, 4, 5

Solutions

UNESCO World Heritage Sites. Using properties of exponential and logarithmic functions

$$(a^n)^m = a^{nm} = a^{mn} = (a^m)^n, b^{\log_b M} = M, \log_{a^n} M^k = \frac{k}{n}\log_a M \text{ and } \log_a a^b = b,$$

let us simplify the expression $\log_{\sqrt[3]{a}} b^{\log_{\sqrt{b}} a^2}$ as

$$\log_{\sqrt[3]{a}} b^{\log_{\sqrt{b}} a^2} = \frac{1}{\frac{1}{3}} \cdot \log_a b^{\frac{2}{1/2} \cdot \log_b a} = 3 \cdot \log_a b^{4 \cdot \log_b a} = 3 \cdot \log_a \left(b^{\log_b a}\right)^4$$
$$= 3 \cdot \log_a a^4 = 3 \cdot 4 = 12.$$

L'Anse aux Meadows National Historic Park and Nahanni National Park in Canada, Galapagos Islands and City of Quito in Ecuador, Simien National Park and Rock-Hewn Churches in Ethiopia, Aachen Cathedral in Germany, Krakow's historic center and Wieliczka and Bochnia Salt Mines in Poland, Island of Goree in Senegal, Mesa Verde and Yellowstone National Parks in the United States were the first 12 landmarks to be inscribed in the UNESCO World Heritage List in 1978.

Countries with the Most World Heritage Sites. The transcendental equation

$$\ln(x^2-56x-57)=\frac{\ln(59-x)}{\ln(x^2-58)}+\ln 59$$

is really challenging to solve. Let us check its domain first

$$\begin{cases} x^2-56x-57>0 \\ 59-x>0 \\ x^2-58>0, x^2-58\neq 1 \end{cases} \Rightarrow \begin{cases} (x+1)(x-57)>0 \\ x<59 \\ x<-\sqrt{58}, x>\sqrt{58}, x\neq \pm\sqrt{59} \end{cases}$$
$$\Rightarrow \begin{cases} x<-1, x>57 \\ x<59 \\ x<-\sqrt{58}, x>\sqrt{58}, x\neq \pm\sqrt{59} \end{cases}.$$

The domain is the interval $57<x<59$ with only one integer $x=58$ inside. Let us verify whether $x=58$ is a solution. Substituting $x=58$ to the equation we get

$$\ln(58^2-56\cdot 58-57)=\frac{\ln(59-58)}{\ln(58^2-58)}+\ln 59 \Rightarrow \ln 59$$
$$=\frac{\ln 1}{\ln(58^2-58)}+\ln 59 \Rightarrow \ln 59 = 0+\ln 59,$$

meaning that $x=58$ satisfies the equation.

Italy has 58 UNESCO World Heritage Sites, more than any other country.

UNESCO Heritage Cities. Using the properties

$$\sqrt[k]{a}=a^{\frac{1}{k}}, (a^k)^m=a^{km}, \text{and} \log_{a^n} M^k=\frac{k}{n}\log_a M,$$

we can simplify the expression as

$$\log_{\frac{1}{8}}\log_4\sqrt{\sqrt[8]{\sqrt[16]{2}}}=\log_{\frac{1}{8}}\log_4 2^{\frac{1}{16\cdot 8\cdot 2}}=\log_{\frac{1}{8}}\log_4 4^{\frac{1}{16\cdot 8\cdot 2\cdot 2}}$$
$$=\log_{8^{-1}}\frac{1}{16\cdot 32}=\log_{8^{-1}}8^{-3}=3.$$

The town of Paphos, the Painted Churches in the Troodos Region, and archeological site Choirokoitia are three UNESCO World Heritage Sites in Cypris. The Great Mosque of Cordoba, Medina Azahara, Alcazar of Cordoba, and Patios Festival are four UNESCO World Heritage Sites located in Cordoba.

Countries with the Least Heritage Sites. Since $2^0 = 1$ and $\log 1 = 0$, the product

$$\log 2^{10} \cdot \log 2^9 \cdot \log 2^8 \cdot \ldots \cdot \log 2^2 \cdot \log 2^1 \cdot \log 2^0 \cdot \log 2^{-1}$$
$$\cdot \log 2^{-2} \cdot \ldots \cdot \log 2^{-9} \cdot \log 2^{-10} \text{ is } 0.$$

Bhutan, Guinea-Bissau, Guyana, Liberia, Sierra Leone, Somalia, and South Sudan do not have any World Heritage Sites.

Restricted UNESCO Heritage Sites. Using the properties of exponential and logarithmic functions

$$\log_{a^n} M^k = \frac{k}{n}\log_a M, \log_a 1 = 0, \log_a M = \frac{1}{\log_M a}, a^0 = 1, \text{ and}$$
$$a^{\log_a M} = M, \text{as } a \neq 1, a > 0, M > 0,$$

the expression $\sqrt[3]{2^{\frac{\log_{81} 1}{\log_9 4}} \cdot 27^{\frac{1}{\log_4 9}}}$ can be simplified as

$$\sqrt[3]{2^{\frac{\log_{81} 1}{\log_9 4}} \cdot 27^{\frac{1}{\log_4 9}}} = \sqrt[3]{2^{\frac{0}{\log_9 4}} \cdot 3^{3\log_3 2}} = \sqrt[3]{1 \cdot 2^3} = 2.$$

Okinoshima Island in Japan, and Mouth Athos in Greece are the only two UNESCO World Heritage Sites restricted to women. Both religious places are open only for men.

Losing Heritage Status. Since

$$\log_a M + \log_a N = \log_a MN, k\log_a M = \log_a M^k, \text{and} \log_a M = \frac{1}{\log_M a},$$

the expression $\log_2 1000 - \frac{1}{\log_5 \sqrt[3]{2}}$ can be simplified as

$$\log_2 10^3 - \frac{1}{\log_5 2^{\frac{1}{3}}} = 3 \cdot \log_2 10 - \frac{1}{\frac{1}{3} \cdot \log_5 2} = 3 \cdot \log_2 10 - 3 \cdot \log_2 5 = 3 \cdot \log_2 2 = 3.$$

Only three sites have been removed from the World Heritage List: the Arabian Oryx Sanctuary of Oman (1994–2007), the Dresden Elbe Valley in Germany (2004–2009), and Liverpool Maritime Mercantile City, England (2004–2021).

Disjointed UNESCO Heritage Sites. Let us consider the first equation $\log_x y = \log_y x$ of the system. The domain is $x > 0, x \neq 1, y > 0, y \neq 1$. Moreover, x and y represent digits of a number, then they must be nonnegative integers. Applying the property $\log_a M = \dfrac{1}{\log_M a}$ of logarithms, the equation can be rewritten as $\log_x y = \dfrac{1}{\log_x y}$ that after the substitution $t = \log_x y$ becomes $t^2 = 1$ that has the solution $t = 1$ and $t = -1$. The back substitution leads to equations $\log_x y = -1$ or $\log_x y = 1$ that are solved as:

- $\log_x y = -1 \Rightarrow x = 1/y$. The equation $x = 1/y$ has only one integer solution $x = 1$ and $y = 1$ that is not in the Domain.
- $\log_x y = 1 \Rightarrow x = y$. Considering $x = y$ with the second equation, we get $3x - 2x = 7$ or $x = y = 7$.

Finally, $2x/y = 2$ and $(9y - 2) = 61$.

The greatest number of sites added to the UNESCO Heritage list in a single year is 61 and the smallest number is 7. Combining 727 different locations of neolithic rock art, the Rock Art of the Mediterranean Basin is the most disjointed World Heritage Site.

Heritage City on a Coral Reef. Using the properties of logarithms,

$$\log_a MN = \log_a M + \log_a N, \log_a \frac{M}{N} = \log_a M - \log_a N \log_a M^k$$
$$= k \log_a M \text{ and } \log_a M = \frac{1}{\log_M a},$$

we can simplify the expressions as

$$\log_3 40.5 = \log_3 \frac{81}{2} = \log_3 81 - \log_3 2 = \log_3 3^4 - \log_3 2 = 4 - \log_3 2$$

$$\log_3 54 = \log_3 2 \cdot 27 = \log_3 2 + \log_3 3^3 = \log_3 2 + 3$$

The sum $4 - \log_3 2$ and $\log_3 2 + 3$ is 7.

The difference between $4 - \log_3 2$ and $\log_3 2 + 3$ is $1 - 2\log_3 2$, which after adding $\log_3 4$ gives 1.

Guam, Kiribati, Marshall Islands, Federated States of Micronesia, Nauru, Northern Mariana Islands, and Palau are seven countries of Micronesia.

Except for the US Commonwealth of the Northern Mariana Islands, Guam and Wake Island, these countries are independent states.

Nan Madol is its only UNESCO World Heritage Site in Federated States of Micronesia.

UNESCO Heritage Lakes. Using the properties of exponential and logarithmic functions $\log_{a^n} M^k = \frac{k}{n}\log_a M$ the expression

$$\log_{125} 5 \cdot \log_{81} 27 = \log_{5^3} 5 \cdot \log_{3^4} 3^3 \text{ can be simplified to } \frac{1}{3} \cdot \frac{3}{4} = \frac{1}{4}.$$

Then 11500·4 = 46000.

The Republic of Malawi takes 46 000 sq mi.

UNESCO Archipelagos. Let us simplify the equation

$$\log_2 \frac{x}{8} = \log_x \frac{1}{4} \text{ at } x > 0, x \neq 1,$$

using the properties of logarithms $\log_{a^n} M^k = \frac{k}{n}\log_a M$ and $\log_a M = \frac{1}{\log_M a}$ as

$$\log_2 x - \log_2 2^3 = -2 \cdot \log_x 2 \Rightarrow \log_2 x - 3 = \frac{-2}{\log_2 x}.$$

Multiplying both sides of the equation by $\log_2 x \neq 0$, we obtain the quadratic equation with respect to $\log_2 x$:

$$\log_2^2 x - 3\log_2 x + 2 = 0 \Rightarrow (\log_2 x - 2)(\log_2 x - 1) = 0$$

$$\Rightarrow \log_2 x = 2 \text{ or } \log_2 x = 1 \Rightarrow x = 4 \text{ or } x = 2.$$

Remark. The substitution $t = \log_2 x$ to $\log_2^2 x - 3\log_2 x + 2 = 0$ can be introduced to simplify the solution. Then the quadratic equation becomes: $t^2 - 3t + 2 = 0$, $t = 2, t = 1$, that after a back substitution leads to the same result.

The Republic of Seychelles consists of four archipelagos, the Republic of the Marshall Islands and the Republic of Mauritius occupy four archipelagos, the Republic of Cabo Verde, the Republic of Vanuatu, and the Republic of Fiji have one archipelago, Solomon Islands contains five archipelagos.

Sources

everything-everywhere.com/interesting-facts-and-stats-about-world-heritage-sites/

www.cnn.com/travel/article/unesco-first-12-world-heritage-sites, kids.kiddle.co/World_Heritage_Site

whc.unesco.org/en/list/

whc.unesco.org/en/statesparties/cy

thefactfile.org/interesting-facts-cyprus/

www.worldatlas.com/articles/city-with-the-highest-number-of-unesco-world-heritage-sites.html

en.wikipedia.org/wiki/Former_UNESCO_World_Heritage_Sites

www.factsinstitute.com/countries/facts-about-micronesia/

en.wikipedia.org/wiki/Micronesia

en.wikipedia.org/wiki/Malawi

africanchildtrust.org.uk/malawifacts/

en.wikipedia.org/wiki/Archipelagic_state

2

Around the World with Equations

> If I were again beginning my studies, I would follow the advice of Plato and start with mathematics... Nature is written in mathematical language.
>
> ~ Galileo Galilei (1564 – 1642)[1]

Breathtaking facts about new inventions and re-discoveries of the past, predictions and reality, history and geography, customs and cultures are revealed while solving the interesting problems of four sections of this chapter. Each problem requires setting a mathematical formulation of real-life phenomena using linear and nonlinear equations, manipulating them to arrive at a solution, and, finally, selecting a realistic answer with suitable applied meaning. The systems of linear and nonlinear equations are arranged by difficulty, beginning with the easiest and progressing to more advanced. While solving these problems, algebra will be reviewed and applied to its core. Interweaving its fundamental relations will not only advance your knowledge and applicability of algebraic identities and rules but will also move you beyond basic mathematical principles toward substantial know-hows through the improvement of problem-solving skills required in many disciplines. Put on your thinking cap and get started!

2.1 Inventions with Linear Equations

Driving on a Causeway

Causeways, bridges, and tunnels have been built to connect islands to continents. They have also significantly influenced the traditions and customs of neighboring countries. The King Fahd Causeway that includes five bridges and causeways was built in 1986 to connect an island country Bahrain with Saudi Arabia. The Bahrainis, like the British, used to drive on the left but changed to driving on the right in 1967 before opening the

[1] Galilean relativity, moons, craters, and thermometers are named after this Italian astronomer, physicist, engineer, and polymath.

DOI: 10.1201/9781003351702-2

causeway to follow the Saudi Arabians that drive on the right. Great Britain and France connected by the underwater railway tunnel are still driving on different sides.

Problem. Half of the length of the King Fahd Causeway is greater than its one-fifth by 7.5 km. How long is the King Fahd Causeway?

Answer. 25 km

Linking to the Continent

The island country of Singapore has been connected to the Southeast Asian country of Malaysia by Johor–Singapore Causeway across the Straits of Johor since 1924. The causeway is one of the busiest border crossings in the world. It also serves as a water pipeline between the two countries. The causeway used to be the only land link between these countries until the Tuas Second Link was open in 1998.

Problem. The third of the length of the Johor–Singapore Causeway is greater than its one-eighth by 0.22 km. How long is the Johor–Singapore Causeway?

Answer. 1.056 km

Temperature Scales

Temperature describes hotness or coldness in numbers. There are three commonly accepted temperature scales, the Fahrenheit (°F) temperature scale mostly used in the United States and a few other English-speaking countries, the Celsius (°C) temperature scale used by all countries that have adopted the metric system of measurement, and the Kelvin (K) or the absolute temperature scale recognized as the international standard for scientific temperature measurement.

The German physicist Daniel Gabriel Fahrenheit proposed a temperature scale based on 32 for the freezing point of water in 1724. The Celsius scale was renamed to honor Anders Celsius in 1948 who developed a similar temperature scale with 0 °C for the freezing point of water in 1742. Scottish William Thomson, the first Baron Kelvin, suggested an absolute thermodynamic temperature scale in 1848. These three temperature scales are related by mathematical relations.

Problem. The Fahrenheit (F), Celsius (C), and Kelvin (K) temperature scales are related as:

$$F = 9/5C + 32^{\circ}, K = C - 273.15^{\circ}.$$

Is it possible to have equal numeric values of temperature in these temperature scales? What values, if yes.

Answer. –40

Nobel Prize

The most prestigious and recognized international award is the Nobel Prize given in Physics, Chemistry, Medicine or Physiology, Literature, and Peace. The founder of Nobel Prizes, Alfred Nobel, was born in Stockholm, Sweden, in 1833 and died in 1896. He was a man of many talents, a great inventor, and generous philanthropist. His 355 patents reflect his contribution to business, chemistry, and engineering. But he is best known for discovering dynamite. Understanding what damage his invention could bring to the society, he left the largest share of his fortune of 31 million Swedish kroner or approximately 300 million US dollars to establish the foundation that was named after him. Each Nobel Prize consists of a gold medal, a diploma bearing a citation, and a sum of money that depends on the income of the Nobel Foundation. Only living people or organizations can be nominated for this prize.

Problem. Find the numbers

- $|S^4_{666}|$ of people who have received the Nobel Prize twice,
- $(S_{2023} + 3)$ of people who have received the Nobel Prize twice in different disciplines,
- S_{444} of people who have received the Nobel Prize three times,
- the maximum number of the Nobel Prizes $2S^2_{777} + 7$ received by one family,

if the sum $S_n = i + i^2 + i^3 + i^4 + i^5 + i^6 + \ldots + i^n$, where n is the number of terms in the sum and $i^2 = -1$. Name these people or those that you know.

Answer. 4, 2, 0, 5

Easter Island

Easter Island is literally in the middle of nowhere, deep in the Pacific Ocean, 3600 km away from the nearest coast of Chile. Its volcanic heights are over 1,600 feet above sea level. Its original inhabitants called the island *Rapa Nui* meaning *Great Rapa* or *Te Pito te Henua* meaning *the Navel of the World* before an expedition led by Dutch admiral Jacob Roggeveen landed on the shore on Easter Sunday of 1722 re-naming the island. This island is not only beautiful but also full of mysteries. Its magnificent 887 iconic stone heads, called the *moai*, have hidden bodies extended into the ground. They were carved between the 12th and 16th centuries. The moai were neither standing up nor in their current locations when first seen by Europeans. It is still an unsolved mystery of how they were able to move.

The Rapa Nui National Park has been inscribed as a UNESCO World Heritage Site since 1995.

Problem. Xavier, Yakov, and Zahari went to explore Easter Island and face its mysteries. During the trip, Xavier paid for food, Yakov paid for logging, Zahari paid for air tickets. At the end of the trip the friends calculated that the amounts each of them paid were x, y, z, $x < y < z$, where x, y, z are the totals paid by Xavier, Yakov, and Zahari correspondingly. How much should Yakov pay or receive so that the friends share the costs equally?

Answer. $\frac{x+z-2y}{3}$

Astonishing Excavations

Petra is an archaeological city in southern Jordan and a symbol of the country. Its name came from Arabic *Al-Batra* or Ancient Greek *Rock* for it carved out of pink sandstone rocks. The city was also known as *Raqemo* after its royal founder. Founded in 312 BC, Petra was the prosperous capital of the powerful Nabataean Kingdom and the center of caravan trade. Its perfect geographical location and constructions of dams, canals, and hydraulic systems to store water created a well-protected artificial oasis and made this desert city grow and flourish. Petra is famous for its astonishing Hellenistic architecture. The palace, treasury, theatre, market, pool, and garden were comfortably located within the city center. The facades of the tombs were decorated with figures of gods and goddesses. Being a royal resting place, Petra has more tombs than the Valley of the Kings in Egypt. Its monastery, used for Christian worship, was carved into the rock. Now it is a holy place for pilgrims. The shrine of the Prophet Aaron's tomb is sacred to Muslims. Petra remained a guarded secret until Swiss Johann Ludwig Burckhardt discovered the city, which still hides its mysterious history.

Tikal in a rainforest of Guatemala is one of the largest archeological ruins of an ancient urban center and the cradle of the pre-Columbian Maya civilization. A Guatemalan newspaper named the discovered place *Tikal* meaning *Place of Voices* in Maya Itza. Although Tikal did not have any water other than stored rainwater, Mayans designed vast causeways, astonishing palaces, towering temples, and enormous pyramids. Tikal was at its peak from the 3rd to 10th century. Its abandonment remains a mystery.

In 1979 Tikal and in 1985 Petra became UNESCO World Heritage Sites.

Problem. Petra was discovered in the year such that three times a year in a century (the number formed by the last two digits of the year) is twice the number formed by the first two digits of the year. The sum of both numbers is 30. Tikal was discovered 36 years later. When were Petra and Tikal discovered?

Answer. 1812, 1848

Rediscoveries of the Past

Two splendid sites have been waiting for centuries before coming back to life. The city Machu Picchu was built atop the Andes Mountain peak of Peru in the 15th century by the Incan ruler Pachacuti and was abandoned when Spaniards conquered the Inca civilization 90 years later. Its name means *Old Peak* or *Old Mountain* in the indigenous Quechua language. The terraced city of Machu Picchu with its 200 religious, ceremonial, agricultural, and astronomical structures was planned and built with highly developed constructing techniques. The indigenous peasant who guided Hiram Bingham to the ruins of this lost Incan city received a payment of one American silver dollar. Machu Picchu was designated a UNESCO World Heritage Site in 1983.

The Rosetta Stone was carved over 2200 years ago to worship the reign of Egyptian pharaoh Ptolemy V Epiphanes who became a ruler at the age of five in 204 BCE. The stone honors his legitimacy, good deeds, and achievements. Its three columns with three different scripts, the sacred script of the empire hieroglyphics that died out in the 4th century, the Egyptian demotic common language, and the official Greek language, helped with deciphering hieroglyphics that puzzled scholars for centuries. The stone was discovered by French soldiers near the city of Rosetta or el-Rashid during their occupation of Egypt. After defeating Napoleon's forces at Alexandria in 1801, the British brought the Rosetta Stone with other Egyptian artifacts to London's British Museum. The museum has declined Egypt's request of returning the stone but gave a full-size replica of it in 2005.

Problem. Find the years when the Rosetta Stone and Machu Picchu (later) were re-discovered if their first two digits of the years are consequent prime numbers with the total sum greater than 30. The last two digits of the year when the Rosetta Stone was found coincide with the largest solution that satisfies

$$|x-59| = 49 - \frac{x}{11},$$

while the lowest solution reveals the last two digits of the year when Machu Picchu was re-discovered.

Answer. 1799, 1911

Unsinkable

Shipwrecks... Even paintings by Rembrandt, Joseph Turner, Claude-Joseph Vernet, and Aivazovsky could not depict the true horror, disaster, and... death. Are there any lucky survivors? Fortunately, yes. Moreover, Fate gave some of them a generous chance to beat the odds and survive not just one but several deadly shipwrecks. An Argentine-English woman of Irish heritage, Violet Jessop, worked as a stewardess and nurse at different ships including sadly known RMS *Titanic* and HMHS *Britannic*. She was onboard these ships during their deadly collisions. And ... she survived. She passed away in 1971 at 83. The 1958 movie *A Night to Remember*, 1997 blockbuster *Titanic*, 1979 movie *SOS Titanic*, 2000 movie *Britannic*, 2020 novel *The Deep*, and others feature some stories of her wonderful life, the protected life of *Miss Unsinkable* or *Queen of Sinking Ships*.

Englishman Arthur John Priest worked as a stoker at different ships and was onboard RMS *Titanic* and HMHS *Britannic* when they sunk. Regrettably, after these and other such sad accidents that he witnessed, nobody wanted to sail with him, and *Mr. Unsinkable* Stoker Arthur John Priest retired. He died in 1937 at the age of 49. He was awarded the Mercantile Marine Ribbon for his service in the Great War.

Problem. Find the numbers of deadly shipwrecks, including *Titanic* and *Britannic*, that Miss Unsinkable Violet Jessop and Mr. Unsinkable Stoker Arthur John Priest survived if these numbers are the lowest for Jessop and the largest for Priest solutions to the equation

$$|3x-10|=|2-x|.$$

Moreover, the year in the 20th century when RMS *Titanic* sank is the product of all solutions to the equation and the year in the 20th century when HMHS *Britannic* sank is the largest solution to the equation squared. Find these sad years.

Answer. 3, 4, 1912, 1916

Predictions, Predictions

The horse is here to stay but the automobile is only a novelty, a fad. Airplanes are interesting toys but of no military value. A rocket will never be able to leave the Earth's atmosphere.

Television won't last because people will soon get tired of staring at a plywood box every night. TVs aren't really going to be a big thing; they are flash in the pan.

Computers in the future may... weigh only 1.5 tons. There is no reason anyone would want a computer in their home...

Are you surprised to read this? These and similar predictions made by leading newspapers, radio pioneers, 20th Century Fox, UK Royal Academy, founder of DEC, and other respectful sources in the 20th century make you smile. Indeed, you enjoy and are not tired of watching a movie on your TV, computer, and phone. And you own not only one computer… Yes, the experts can miss the mark when it comes to forecasting the future… Fast development of technological innovations change our surrounding extremely fast.

Problem. The solutions arranged from the lowest to the largest value to the following equation

$$|2x-169|+2x=|59-x|+157$$

unveil the last two digits of the year in the 20th century when the first tape recorder, the first email, and the first www were created. Do you know the names of these out-of-box inventors? What does the abbreviation www mean?

Answer. 1947, 1971, 1989

Credit Cards

Credit cards have become a part of our life since businessman Frank McNamara introduced it in 1950. When he tested a suggested several-digit PIN code to his wife, she said that she as well as any other woman could not remember such a long sequence of numbers and proposed her number of digits.

Problem. The proposed numbers of PIN code digits by Frank McNamara and his wife are the largest and lowest values of the solutions to the following equation:

$$|x-5|+|x-4|=|x-3|$$

Find these numbers.

Answer. 4, 6

Solutions

Driving on a Causeway. Let x km be the length of the King Fahd Causeway. Then

$$\frac{x}{2}-\frac{x}{5}=7.5 \Rightarrow \frac{5x-2x}{10}=7.5 \Rightarrow x=25.$$

The King Fahd Causeway that connects two kingdoms, Bahrain and Saudi Arabia, is 25 km.

Linking to the Continent. Let x km be the length of the Johor–Singapore Causeway. Then

$$\frac{x}{3}-\frac{x}{8}=0.22 \Rightarrow \frac{8x-3x}{24}=0.22 \Rightarrow x=1.056$$

The Johor–Singapore Causeway that connects Singapore and Malaysia is 1.056 km or 0.66 mi.

Temperature Scales. Using the conversion formula for a temperature expressed in the Celsius (C) scale to its Fahrenheit (F) representation, we get

$$F = 9/5C + 32^{\circ} = C \Rightarrow 4/5C = -32^{\circ} \Rightarrow C = -40^{\circ}.$$

Using the conversion formula for a temperature expressed in the Celsius (C) scale to its Kelvin (K) representation, we obtain

$$K = C - 273.15^{\circ} = C \Rightarrow -273.15^{\circ} = 0^{\circ}, \text{ impossible.}$$

That is, –40 °C = –40 °F. Numeric values of temperatures in Kelvin and Fahrenheit or Celsius cannot be the same.

Nobel Prize. Remembering that $i^2=-1, i^3=-i, i^2=1$, and $i^{n+1}=i \cdot i^n$, we can rewrite the sum $S_n = i+i^2+i^3+i^4+i^5+i^6+...+i^n$ as $S_n = i-1-i+1+i-1-i+1+...+i^n$ and note that the pattern $i-1-i+1$ repeats. Therefore, depending on n, the answer will be

$$S_n = \begin{cases} 0, & n=4k, \\ i, & n=4k+1, \\ i-1, & n=4k+2, \\ -1 & n=4k+3. \end{cases}$$

S_{666} is the sum of 666 terms. On the other side, $n = 666 = 4 \cdot 166 + 2$, where $k = 166$. Thus, $S_{666} = S_2 = i - 1$. Finally, $|S^4_{666}| = |(i-1)^4| = 4$.

Similarly, for $n = 2023 = 4 \cdot 505 + 3 = 4 \cdot k + 3$, where $k = 505$. Then $S_{2023} + 3 = S_3 + 3 = -1 + 3 = 2$.

For $n = 444 = 4 \cdot k$, where $k = 111$, $S_{444} = 0$.

Finally, for $n = 777 = 4 \cdot 194 + 1 = 4 \cdot k + 1$, where $k = 194$, we get $2S^2_{777} + 7 = 2S^2_1 + 7 = 2 \cdot i^2 + 7 = 2 \cdot (-1) + 7 = 5$.

There are no people who have been awarded the Nobel Prize three times, though the International Committee of the Red Cross was awarded the Nobel Peace Prize three times, in 1917, 1944, and 1963.

Polish-French scientist Marie Curie with the 1903 Nobel Prize in Physics and the 1911 Nobel Prize in Chemistry, US chemist Linus Pauling with the 1954 Nobel Prize in Chemistry and the 1962 Nobel Peace Prize, US engineer John Bardeen with the 1956 and 1972 Nobel Prizes in Physics, and Chemist Frederick Sanger with the 1958 and 1980 Nobel Prize in Chemistry are the only four people awarded the Nobel Price twice. The first two scientists are the only ones that received this prestigious award in different disciplines.

Four members of the Curie family, Pierre Curie, his wife Marie Curie (twice), their daughter Irène, and son-in-law Frédéric Joliot-Curie, were awarded five Nobel Prizes.

Easter Island. The average of amounts each of them paid is $A = \frac{x+y+z}{3}$. Then $\frac{x+y+z}{3} - y = \frac{x+z-2y}{3}$ shows how much Yakov should receive or pay.

- If $\frac{x+z-2y}{3} > 0$, then Zahari paid $\frac{x+z-2y}{3}$ to Yakov.
- If $\frac{x+z-2y}{3} = 0$, then Yakov neither received nor paid any amount.
- If $\frac{x+z-2y}{3} < 0$, then Yakov paid the amount of $\frac{x+z-2y}{3}$ to Xavier.

Astonishing Excavations. Let y be the year in a century, then the century is $3y/2$ and $y + 3y/2 = 30$, from which follows $y = 12$.

Pretending to be a Bedouin, Johann Ludwig Burckhardt discovered Petra in 1812 during his expedition in the Transjordan. Tikal was founded in 1848.

Rediscoveries of the Past. It is easy to see that the first two digits of the years are consequent prime numbers with the total sum greater than 30 are 17 and 19.

The absolute value in the equation $|x-59| = 49 - \frac{x}{11}$ can be split to

$$|x-59| = \begin{cases} 59-x, & x < 59, \\ x-59, & x \geq 59. \end{cases}$$

Hence, if x < 59, then the equation

$$|x-59| = 49 - \frac{x}{11} \text{ becomes } 59 - x = 49 - \frac{x}{11},$$

that has the solution $x = 11 < 59$. If $x \geq 59$, then it becomes

$$x - 59 = 49 - \frac{x}{11}, \text{ that leads to } x = 99 \geq 59.$$

The Rosetta Stone was found by French soldiers during their occupation of Egypt in 1799. Machu Picchu was re-discovered by Yale Professor Hiram Bingham in 1911.

Unsinkable. The absolute values in the equation $|3x-10| = |2-x|$ should be treated differently at three intervals:

1. $x < 2$, then the equation $|3x-10| = |2-x|$ becomes $10 - 3x = 2 - x$ that leads to $x = 3$, which is greater than 2. Thus, there is no solution on the interval $(-\infty, 2)$.
2. $2 \leq x \leq 10/3$, then $3x - 10 = 2 - x$ and $x = 3$. Since $2 \leq 3 \leq 10/3$, $x = 3$ is the solution on the interval [2, 10/3], and, thus, a solution to the equation $|3x-10| = |2-x|$.
3. $x > 10/3$, then $3x - 10 = x - 2$ and, then, $x = 4$. Since $4 > 10/3$, and it is a solution to the equation $|3x-10| = |2-x|$.

Miss Unsinkable Violet Jessop survived three deadly shipwrecks, RMS *Olympic* when it was collided with the British warship HMS *Hawke* in 1911, RMS *Titanic* in 1912, and their sister ship HMHS *Britannic* in 1916.

Mr. Unsinkable Stoker Arthur John Priest survived four ship sinkings, RMS *Titanic*, HMS *Alcantara*, HMHS *Britannic*, and SS *Donegal*.

Predictions, Predictions. Let us consider our equation on three intervals and present our findings in the table.

Interval	$x \leq 59$	$59 < x < 84.5$	$x \geq 84.5$
$\lvert 2x-169 \rvert$	$169 - 2x$	$169 - 2x$	$2x - 169$
$\lvert 59-x \rvert$	$59 - x$	$x - 59$	$x - 59$
$\lvert 2x-169 \rvert + 2x = \lvert 59-x \rvert + 157$	$169 - 2x + 2x = 59 - x + 157$	$169 - 2x + 2x = x - 59 + 157$	$2x - 169 + 2x = x - 59 + 157$
x	47	71	89

The first tape recorder appeared in 1947. American Ray Tomlinson invented electronic mail in 1971. British Tim Berners-Lee designed the Word Wide Web in 1989 to automatically share information between scientists around the globe.

Credit Cards. Remembering that

$$|x-a| = \begin{cases} a-x, & \text{if } x \le a \\ x-a, & \text{if } x > a \end{cases},$$

let us consider our equation on four intervals and present our findings in the table, where the first row shows the considered interval, and the second row shows the transformed equation.

$x \le 3$	$3 < x \le 4$	$4 < x < 5$	$x \ge 5$
$5-x+4-x=3-x$ $\Rightarrow x=6>3$ $\Rightarrow x=6$ is not a solution	$5-x+4-x=x-3 \Rightarrow x=4$ $3<4\le 4 \Rightarrow x=4$ is a solution	$5-x+x-4=x-3$ $\Rightarrow x=4$ 4 is not between 3 and 4 $\Rightarrow x=4$ is not a solution on this interval	$x-5+x-4=x-3$ $\Rightarrow x=6$ $6\ge 5 \Rightarrow x=6$ is a solution

Frank McNamara suggested 6-digit PINs, but his wife convinced him to go with the 4 digits that we currently use.

Sources

headstuff.org/culture/history/terrible-people-from-history/margaret-brown-unsinkable-lady-titanic/

www.weforum.org/agenda/2016/10/10-predictions-for-the-future-that-got-it-wildly-wrong

www.rd.com/list/predictions-that-were-wrong/

www.scoopwhoop.com/news/interesting-facts-about-the-nobel-prize-/

www.nobelprize.org/prizes/themes/the-red-cross-three-time-recipient-of-the-peace-prize/

phys.org/news/2022-10-scientists-won-nobel-prizes.html, en.wikipedia.org/wiki/Curie_family

www.livescience.com/temperature.html

en.wikipedia.org/wiki/Johor%E2%80%93Singapore_Causeway

en.wikipedia.org/wiki/List_of_island_countries

en.wikipedia.org/wiki/King_Fahd_Causeway

en.wikipedia.org/wiki/Petra

www.wanderlustchloe.com/facts-about-petra/
theculturetrip.com/central-america/guatemala/articles/10-fascinating-things-you-didnt-know-about-tikal-maya-ruins
en.wikipedia.org/wiki/Tikal
www.atlasandboots.com/travel-blog/easter-island-facts/
factcity.com/facts-about-easter-island/
www.factretriever.com/machu-picchu-facts
learnodo-newtonic.com/the-rosetta-stone-facts
www.mentalfloss.com/article/82799/15-solid-facts-about-rosetta-stone

2.2 Nonlinearities around the Globe

The Cradle of Civilization

The Republic of Iraq in Western Asia has amusing history and iconic heritage. Its name meaning *deeply rooted, well-watered, and fertile* in Arabic has been used since the 6th century. Iraq borders with Turkey, Iran, Kuwait, Saudi Arabia, Jordan, and Syria. Its location between the Tigris and Euphrates, known as *Mesopotamia*, is recognized as the birth of the oldest civilization that shaped the first empires, cities, and laws, and created mathematics, timekeeping, a calendar, astrology, writing, and cuneiform script on clay tablets. The region was conquered many times. Finally, the Hashemite Kingdom of Iraq gained independence from the UK. The monarchy was overthrown a few decades later giving the birth to the Iraqi Republic.

The country's capital city, Baghdad, has always been the largest multicultural city and appeared in *Arabian Nights* and many other books, legends, and folk stories. Islamic books state that the biblical Noah's ark was built in the Iraqi city of Kufa.

It is interesting to mention that the superstition against black cats began in Iraq.

Problem. The fourth digit of the year when the Hashemite Kingdom of Iraq gained independence from the UK is the value of n for which all solutions to the quadratic equation

$$nx^2 + 2(n-8)x + (3n+12) = 0$$

are equal. The solution to the equation is the third digit of the year. In 26 years, the monarchy was overthrown, and the Iraqi Republic was formed. What years in the 20th century did the Hashemite Kingdom of Iraq gain its independence? When did the Iraqi Republic appear?

Answer. 1932, 1958

Monarchies and Kingdoms

A monarchy is a country led by a single ruler. Monarchies have existed since the first civilizations. Their monarchs have respected as gods. The monarch, also called the king, queen, sultan, emperor, and emir, can be the head of a state, the head of government, or both. There are several kinds of monarchy: an absolute monarchy, where the monarch rules with unlimited power and a constitutional monarchy with an elected government that runs the country. Monarchs generally reign for life and most of them are hereditary, though, some monarchs can be elected like in Malaysia and the Holy Roman Empire. A monarchy is commonly called a kingdom, though there are other names, like the Sultanate of Oman and Grand Duchy of Luxembourg.

The word *monarchy* comes from the combination of Greek *monarkhia* as *monos* meaning *only* and *arkhe* meaning *authority*.

It is interesting that there are 44 monarchies but only 29 monarchs. Asia is leading the list of monarchies.

Problem. The number of kingdoms in Africa is a prime solution to the equation

$$36x^4 - 555x^3 + 1522x^2 - 555x + 36 = 0.$$

The number of kingdoms in Europe is a composite solution to that equation. How many monarchical states are in Africa and Europe? List those that you know.

Answer. 3, 12

The Largest Kingdom

Do you know the largest kingdom? The royal Windsor family used to rule not only over the United Kingdom but also over other countries scattered across three continents. Although these countries are now independent sovereign nations, they still maintain at least a ceremonial connection to the British monarchy. The Commonwealth countries are voluntarily united and can leave it at any time as the Republic of Ireland did in 1949 and Zimbabwe did in 2003. King Charles III succeeded his mother, Elizabeth II, as monarch and became the head of the Commonwealth on September 8, 2022.

Problem. Find the number of countries that consider King Charles III as their king if this number satisfies the equation

$$x^2 - 246 = 2 \cdot \sqrt{x^2 - 231}.$$

List these countries.

Answer. 16

Countries with Similar Names

Nigeria and Niger, Mauritania and Mauritius, Slovakia and Slovenia, Angola and Anguilla, Dominica and Dominican Republic, Gambia and Zambia, Liberia and Libya, Ecuador and El Salvador, Monaco and Morocco sound so close that one may think that they are the same countries, though some of them are far away from each other. Mauritania is a country in Northwest Africa, while Mauritius is an Indian Ocean Island nation. It is necessary to cross the Mediterranean Sea to go from tiny European city-state Monaco to North African country Morocco. Australia and Austria are located at different continents. At the same time, Iraq and Iran as well as Nigeria and Niger are bordering countries.

Problem. The length in miles of the border between Niger and Nigeria can be presented with three segments of numbers written side by side. The first segment is x^2, the second one is $3x$, and the third segment is the average of the first two numbers, where x satisfies the equation

$$(x+6)^2 + (x+6)^{\frac{1}{2}} = 84.$$

How long is the border between Niger and Nigeria?

Answer. 999 mi

Borders between Countries

The Sahel region between the Sahara Desert to the north and the Sudan Savanna to the south has unique biodiversity. Landlocked Republic of Chad and its bordering Libya, Sudan, the Central African, Cameroon, Nigeria, and Niger are in the Sahel region at least partially. The mostly straight-line border between Chad and Niger is special because it runs from one tripoint with Libya to another tripoint with Nigeria. Tracing mostly within the Sahara Desert, the border does not go close to any small village or road.

The freshwater Lake Chad is the jewel of the border and used to be the land of powerful African kingdoms.

Problem. Find the length in kilometers of the border between Niger and Nigeria if it can be presented as $100x$, where a positive integer solution x satisfies the equation

$$(x-10)(x^2+10)=308.$$

Answer. 1200 km

The Land of Eagles

Albania is a country on Southeastern European Balkan Peninsula with Adriatic and Ionian coastlines and the Albanian Alps. This small country is proud of its exceptional biodiversity for it has one third of all plant species found in Europe and many animal endangered species. Lake Ohrid, the oldest lake in Europe, and its unique surrounding and living fossils are protected by UNESCO. Ancient Albanian city Butrint is also among UNESCO World Heritage Sites. The country's oil reserves are the largest in Europe.

Albanians call themselves *shqiptarë* meaning *Sons of the Eagles* or *Land of the Eagles*. There are more Albanians living outside the country than living inside it. Although a 2011 census shows that about 60% of Albanians are Muslims, 17% are Christian, and only 17% are atheist, Albania was proclaimed as the first atheist state in the world. The most famous and the only Albanian to win a Nobel prize is Anjezë Gonxhe Bojaxhiu, better known as Mother Teresa. The Albanian international airport in Tirana is named in her honor.

It is interesting that one of the oldest cities in Europe, Shkodra in Albania, does not have any traffic lights.

Problem. Find the year when Albania became the first atheist state if all integer factors (but 1 and a number itself) of the year are integer solutions x and y of the nonlinear equation

$$xy-2x+2y=1419.$$

The smallest integer that satisfies this nonlinear equation provides the number of UNESCO World Heritage Sites in Albania. Find this number.

Answer. 1967, 3

Lucky–Unlucky Numbers around the World

Numbers… How much magic, mysticism, and symbolism they possess! How much religious, philosophic, and spiritual interpretations they have created since the origin of humanity… Countless predictions and spells have been made based on numbers. Civilizations have been too superstitious about numbers, having a love–hate relationship with them and developing different meanings for them. You must be careful with lucky–unlucky numbers before taking your next trip.

For instance, you bring a dozen of beautiful roses to any party in the Western world to please a host but be careful to bring the same bouquet to a Russian party. The best thing the host can do to avoid ruining the party is to place them in two different vases with an odd number of flowers in each. Indeed, it is a celebration, not a funeral! An even number of flowers is only for such a sad event…

If you are in China, you will see numerous neon signs with 666. No, it is not the Number of the Beast as in the Christian world, it means *Everything goes smoothly* and that everything runs well there.

Problem. Four numbers, lucky in some countries and unlucky in others, arranged from smallest to largest, satisfy the following equation

$$x^4 - 41x^3 + 579x^2 - 3271x + 6188 = 0.$$

The first number is lucky in Germany but unlucky in China, Japan, Vietnam, any other Asian countries.

The second number is lucky in the UK, the USA, France, Netherland, and other Western countries though considered to be unlucky in China, Vietnam, Thailand, or other Asian countries. The third number is unlucky for most of the world but lucky in Italy, while the last one is unlucky there. Find these superstitious numbers. Explain why they have these meanings in different countries.

Answer. 4, 7, 13, 17

Problem. Find the length in kilometers of the border between Niger and Nigeria if it can be presented as $100x$, where a positive integer solution x satisfies the equation

$$(x - 10)(x^2 + 10) = 308.$$

Answer. 1200 km

Lucky–Unlucky Numbers in Countries

Have you travelled to different countries around the world or used international means of transportation? Have you noticed that the deck 17 is missing in Italian MSC cruise line, while the deck 13 is missing in American cruise ships. There is no floor 4 in buildings in Japan and China and you will go from floor 3 to floor 5 at once. Due to their negative associations, these numbers cannot be found among the numbers of airline seats, flights, and parking lots. Humanity has advanced in various ways developing different languages and their own superstitions on numbers. The same numbers can be lucky in one country and unlucky in another.

Problem. The smallest number that satisfies the equation

$$\frac{2}{\sqrt[6]{x^5}}+\frac{1}{3}=\frac{1}{\sqrt{x}}+\frac{2}{3\sqrt[3]{x}},$$

is lucky for Asians but unlucky for Indians and Bulgarians, while the largest one is lucky for Chinese and Norwegians but unlucky to Japanese. Find these numbers. Explain why they have different meanings.

Answer. 8, 9

Speed Limits

Drivers dream to drive as fast as they can, but... speed limits. It is hard to imagine roads and highways without speed limits. Roads cutting through Poland, Texas, and the United Arab Emirates have the highest posted speed limit in the world. Germany's iconic autobahn is famous for its limitless stretches of tarmac, where a recommended limit of 130 km/h is often ignored.

Problem. Find the maximum posted speed limit on some roads in Australia's Northern Territory and the Isle of Man and if it is a positive integer that satisfies the following equation

$$15x^4+16x^3+90x^2+112x-105=0.$$

Answer. no

From Rejection to Fame

The popularity of a series of seven Harry Potter fantasy novels written by British author J. K. Rowling is incredible. The books have been translated into 85 languages. Millions of copies have been sold worldwide since the release of the first book in 1997. This best-selling book series has broken all records as the fastest-selling books in history. Seven books were adapted into an eight-part film series by Warner Bros. These films have also become a huge commercial success.

Do you know that publishing Harry Potter met many obstacles? It was not an immediate success. Rowling had received several rejection letters before the publisher Bloomsbury agreed to publish it though with little enthusiasm and expectation. Her persistence is a perfect example of *Never give up your dreams*.

Problem. The number of times Harry Potter was rejected by different publishers can be presented as the sum of a square and a cube of one number or as the sum of the following number and its square. How many rejection letters did J.K. Rowling receive?

Answer. 12

Solutions

The Cradle of Civilization. The roots of a quadratic equation $ax^2+2bx+c=0$ are equal if its discriminant $b^2-ac$is zero that is,

$$(n-8)^2-n(3n+12)=0 \Rightarrow n^2+14n-32=0 \Rightarrow n=-16 \text{ or } n=2.$$

Next, $n=-16$ cannot be a digit of a year. So, the fourth digit of the year is 2. Finally, 3 satisfies $x^2-6x+9=0$ obtained after substituting 2 to the original equation

$$nx^2+2(n-8)x+(3n+12)=0.$$

Conclusively, 1932 + 26 = 1958.

The Hashemite Kingdom of Iraq gained independence from the UK in 1932. The Iraqi Republic was established in 1958.

Monarchies and Kingdoms. The nonlinear equation $36x^4-555x^3+1522x^2-555x+36=0$ is an equation of the fourth degree, which is challenging to solve. Let us use its wonderful feature of symmetric coefficients, namely, 36, –555, 1522, –555, 36. It is obvious that $x=0$ does not satisfy the equation that allows us to divide both parts of the equation by x^2

$$36x^2-555x+1522-555\cdot\frac{1}{x}+36\cdot\frac{1}{x^2}=0, \text{or } 36x^2-555x+1450+2\cdot 36$$
$$-555\cdot\frac{1}{x}+36\cdot\frac{1}{x^2}=0.$$

Let us introduce a substitution

$$t = x + \frac{1}{x}, \text{then } t^2 = x^2 + 2 + \frac{1}{x^2},$$

and rewrite the equation in terms of t as

$$36t^2 - 555t + 1450 = 0,$$

that, following the quadratic formula

$$t = \frac{555 \pm \sqrt{555^2 - 4 \cdot 36 \cdot 1450}}{2 \cdot 36}, \text{ has solutions } t = \frac{145}{12} \text{ and } t = \frac{145}{12} t = \frac{10}{3}.$$

A backward substitution

$$t = x + \frac{1}{x} = \frac{145}{12} \text{ and} t = \frac{10}{3} t = x + \frac{1}{x} = \frac{10}{3},$$

leads to two quadratic equations $12x^2 - 145x + 12 = 0$ and $3x^2 - 10x + 3 = 0$, with the solutions $x = 12, 1/12, 3, 1/3$. Only integer solutions are of our interest. In addition, 3 is prime and 12 is a composite number.

Kingdom of Morocco, Kingdom of Lesotho, and the Kingdom of Eswatini (formerly Swaziland) are Africa's last three monarchies, while some other African countries are sub-national monarchies.

Of the 12 monarchies in Europe, Denmark, Norway, Sweden, the United Kingdom, Spain, the Netherlands, and Belgium are seven kingdoms, while Andorra, Liechtenstein, and Monaco are principalities, Luxembourg is a Grand Duchy, and Vatican City is a theocratic elective monarchy ruled by the Pope.

The Largest Kingdom. 1st way. Let us transform the equation

$$x^2 - 246 = 2 \cdot \sqrt{x^2 - 231} \text{ to } x^2 - 231 - 15 = 2 \cdot \sqrt{x^2 - 231},$$

use the substitution$t = \sqrt{x^2 - 231}, t \geq 0$, and rewrite the equation in terms of t as $t^2 - 2t - 15 = 0$, or $(t - 5)(t + 3) = 0$. Then $t = 5$, $t = -3$. Only $t = 5 \geq 0$ works. Hence, $5 = \sqrt{x^2 - 231}$ and $x = 16$, $x = -16$. The solution $x = 16$ satisfies the problem.

2nd way. Squaring both sides of the equation$x^2 - 246 = 2 \cdot \sqrt{x^2 - 231}$, leads to the equation of the fourth degree $x^4 - 2 \cdot 246x^2 + 246^2 = 4x^2 - 4 \cdot 231$ that does not have any odd powers of x. The substitution $t = x^2, t \geq 0$, turns this equation into a quadratic equation leading to the same result as above.

The United Kingdom, Antigua and Barbuda, Australia, Bahamas, Barbados, Belize, Canada, Grenada, Jamaica, New Zealand, Papua New Guinea, Saint Kitts and Nevis, Saint Lucia, Saint Vincent and the Grenadines, Solomon Islands, ans Tuvalu are 16 countries that consider King Charles III as their king.

Countries with Similar Names. The domain of the equation

$$(x+6)^2+(x+6)^{\frac{1}{2}}=84 \text{ is } [-6,\infty).$$

Let us consider the interval $[0,\infty)$ first. On this interval the function $y=(x+6)^2+(x+6)^{\frac{1}{2}}-84$ is strictly increasing. In addition $y(0) = -84$ and $y \to \infty$ as $x \to \infty$. Thus, the equation $(x+6)^2+(x+6)^{\frac{1}{2}}=84$ has just one solution. It is easy to see that this solution is 3.

1st way. On the interval $[-6, 0)$, the term $(x+6)^2$ takes values between 0 and 6 and $(x+6)^{\frac{1}{2}}$ between 0 and $\sqrt{6}$. Thus, their sum cannot reach 64, and there is no solution on $(-6, 0)$. There are no negative solutions.

Then $x^2 = 9$, $3x = 9$, and the average is 9.

There is a 999-mile border between Niger and Nigeria.

Borders between Countries. Let us rewrite the original equation

$$(x-10)(x^2+10)=308 \text{ as } x^3-10x^2+10x-408=0.$$

Since the term $(-10x^2 + 10x)$ ends in 0, the last digit of $(x^3 - 408)$ should also be 0, which is possible if the last digit of x^3 is

- 8 if x is positive or
- 2 if x is negative.

Following the problem only positive integer solutions are of our interest. Then the last digit of a solution x to the equation can be 2 or 8 and a solution must be a factor of 408. Only 2, 8, 12, 68, 102, and 408 satisfy these conditions. Checking a few of these numbers, we can get that $x = 12$ is one of the solutions. Hence, $x^3-10x^2+10x-408=0$ can be presented as $(x-12)(x^2+2x+34)=0$. The equation $x^2+2x+34=0$ does not have any real roots because its discriminant $22 - 4{\cdot}34 < 0$. Thus, $x = 12$ is the only solution to $(x-10)(x^2+10)=308$.

Other ways to solve the equation $(x-10)(x^2+10)=308$ is to use a formula for a cubic equation or closely examine pairs of multiples (1, 308), (2, 154), (4, 77), (7, 44), (11, 28), and (14, 22) of 308.

The Chad–Niger border is around 1200 km long.

The Land of Eagles. Let us rewrite the nonlinear equation

$$xy - 2x + 2y = 1419 \text{ as } xy - 2x + 2y - 4 = 1415$$

and regroup terms in the left side as $(x + 2)(y - 2) = 1415$. Since 1415 has four integer factors 1, 5, 283, and 1415, the integer multipliers of the last equation can take only these values. Namely

$$\begin{cases} x+2=1 \\ y-2=1415 \end{cases} \text{or} \begin{cases} x+2=1415 \\ y-2=1 \end{cases} \text{or} \begin{cases} x+2=5 \\ y-2=283 \end{cases} \text{or} \begin{cases} x+2=283 \\ y-2=5 \end{cases}.$$

The first system does not work because then $x = -1$, the second and third systems do not produce a solution because their roots $x = 1413$ and $y = 3$ or $x = 3$ and $y = 285$ lead to unrealistic years of 4239 and 855. The solution to the fourth system $x = 281$ and $y = 7$ gives the year of 1967.

Albania was proclaimed as the first atheist state in the world in 1967. Albania is proud of its three UNESCO World Heritage Sites.

Lucky–Unlucky Numbers around the World. The equation

$$p(x) = x^4 - 41x^3 + 579x^2 - 3271x + 6188 = 0$$

of the fourth degree is not easy to solve. Therefore, let us analyze it first. The equation $p(x) = 0$ has four roots. Since the sign of coefficients of $p(x)$ changes four times, there are 4, 2, or 0 positive real solutions. Following the problem, only integer solutions are of interest to us. They should be also factors of 6188. It is easy to see that 6188 is divisible by 4. Indeed, the number formed by its last two digits 88 is divisible by 4. Let us check whether 4 is one of the solutions to the equation $p(x) = 0$:

$$\begin{array}{cccccc} 4 & 1 & -41 & 579 & -3271 & 6188 \\ & \downarrow & 4 & -148 & 1724 & 6188 \\ & 1 & -37 & 431 & -1547 & 0 \end{array}$$

Yes, indeed. Thus, the original equation $p(x) = x^4 - 41x^3 + 579x^2 - 3271x + 6188 = 0$ can be rewritten as $p(x) = (x-4)(x^3 - 37x^2 + 431x - 1547) = 0$. The digits of the coefficient 1547 satisfy the property of divisibility by 7. Indeed, $154 - 7 \cdot 2 = 140$ is divisible by 7. Using synthetic division again, let us check whether 7 is the solution to $x^3 - 37x^2 + 431x - 1547 = 0$:

$$\begin{array}{ccccc} 7 & 1 & -37 & 431 & -1547 \\ & \downarrow & 7 & -210 & 1547 \\ & 1 & -30 & 221 & 0 \end{array}$$

Hence, the original equation $p(x) = 0$ can be rewritten as $p(x) = (x-4)(x-7)(x^2-30x+221) = 0$. Using the quadratic formula to solve the equation $x^2-30x+221=0$, we can find the last two solutions $x = 13$ and $x = 17$ of the original equation $p(x) = 0$.

The pronunciations of the number 4 and the word for *death* are similar in Chinese and Japanese.

The number 7 is a lucky number is the Western world: God made the universe in 7 days. There are 7 deadly sins, 7 wonders of the ancient world, 7 planets of the ancient world and so on. In Asian countries the seventh month is a month for the *ghost*.

The number 13 is a lucky number in Italy because it is the number of St. Anthony, the patron saint of finding lost people there, but an unlucky number in most Western countries and Americas where 13 is connected to many grim events recorded in the Bible and the Mayan Calendar.

The number 17 is as unlucky in Italy as 13 is to Americans because its Roman writing XVII rearranged to VIXI is translated as *I'm now dead* or *my life is over*.

Lucky–Unlucky Numbers in Countries. Let us rewrite the original equation as

$$\frac{2}{\sqrt[6]{x^5}}+\frac{1}{3}=\frac{1}{\sqrt{x}}+\frac{2}{3\sqrt[3]{x}} \Rightarrow \frac{2}{\sqrt[3]{x}\cdot\sqrt{x}}-\frac{2}{3\sqrt[3]{x}}-\frac{1}{\sqrt{x}}+\frac{1}{3}=0 \Rightarrow$$

$$\frac{2}{\sqrt[3]{x}}\left(\frac{1}{\sqrt{x}}-\frac{1}{3}\right)-\left(\frac{1}{\sqrt{x}}-\frac{1}{3}\right)=0 \Rightarrow \left(\frac{1}{\sqrt{x}}-\frac{1}{3}\right)\cdot\left(\frac{2}{\sqrt[3]{x}}-1\right)=0 \Rightarrow \frac{1}{\sqrt{x}}=\frac{1}{3} \text{ or } \frac{2}{\sqrt[3]{x}}=1,$$

from which follows $x = 9$ or $x = 8$.

The number 8 is lucky for Asians where it sounds similar to *prosper* and *make money*. Even the Summer Olympics of August 2008 held in Beijing started at 08:08:08 local time. However, the number is unlucky for India where it is the number of Sani and related to breaker and lethargic kind of personality.

The number 9 stands for the emperor in China and appears in Norwegian mythology and is lucky there. but unlucky to Japanese, where it sounds like its word *suffering*.

Speed Limits. The equation

$$p(x) = 15x^4+16x^3+90x^2+112x-105=0$$

is a polynomial equation of the fourth degree with real coefficients. Instead of solving it and finding its positive integer solution let us analyze it first. The equation has four roots. All four roots can be complex or real, or two roots can be real then the other two are complex. The signs of real coefficients of $p(x) = 0$ alternate only once. Indeed, they are ++++−. Therefore, there is one positive real solution to $p(x) = 0$, and consequently, there is at least one negative real root or, precisely, one or three negative real roots are possible. Another way to come to this conclusion is to check the signs of real coefficients of $p(-x) = 0$ that alternate three times (+− +−−). Only positive integer solutions are of our interest. They can be presented as a ratio between factors of 105 and factors of 15. Let us check one of them, $x = 15$ and provide a synthetic division of $p(x)$ by a linear factor $(x - 15)$:

$$\begin{array}{rrrrrr} 15 & 15 & 16 & 90 & 112 & -105 \\ & & 225 & 3615 & 55575 & 835305. \\ & 15 & 241 & 3705 & 55687 & 835200 \end{array}$$

There are no positive roots greater than 15 because the last row contains only positive numbers. Let us check $x = 1$ by performing a synthetic division of $p(x)$ by a linear factor $(x - 1)$:

$$\begin{array}{rrrrrr} 1 & 15 & 16 & 90 & 112 & -105 \\ & & 15 & 31 & 121 & 233\ . \\ & 15 & 31 & 121 & 233 & 128 \end{array}$$

The last row contains only positive numbers, then the only positive real root is between 0 and 1 and the polynomial equation $p(x) = 0$ does not have and positive integer solutions.

There is no posted maximum speed on some ridiculously long, straight, flat roads in colossal Australia's Northern Territory and the Isle of Man, which is the self-governing British Crown Dependency in the Irish Sea.

From Rejection to Fame. Let n be the number of rejection letters received by J. K. Rowling, then $n + 1$ is the following number and:

$$\begin{aligned} n^2 + n^3 = (n+1) + (n+1)^2 &\Rightarrow n^2\,(n+1) = (n+1) + (n+1)^2 \\ &\Rightarrow (n+1)\,(n^2 - n - 2) = 0 \\ &\Rightarrow (n+1)\,(n-2)\,(n+1) = 0. \end{aligned}$$

The last equation has just one positive integer solution $n = 2$.

J. K. Rowling received 12 rejection letters for her submission of her Harry Potter novel from 12 different publishing houses.

Sources

en.wikipedia.org/wiki/Iraq, www.muslimaid.org/media-centre/blog/8-surprising-facts-about-iraq/

www.infoplease.com/world/social-statistics/kingdoms-and-monarchs-world

en.wikipedia.org/wiki/Monarchies_in_Africa, en.wikipedia.org/wiki/Monarchies_in_Europe

worldpopulationreview.com/country-rankings/countries-with-royal-families

en.wikipedia.org/wiki/Commonwealth_realm

en.wikipedia.org/wiki/Chad%E2%80%93Niger_border

www.britannica.com/place/Lake-Chad

en.wikipedia.org/wiki/Niger%E2%80%93Nigeria_border

sottyreview.wordpress.com/2015/01/04/lucky-and-unlucky-numbers-from-around-the-world/

people.howstuffworks.com/13-superstitions-about-numbers.htm

sottyreview.wordpress.com/2015/01/04/lucky-and-unlucky-numbers-from-around-the-world/ people.howstuffworks.com/13-superstitions-about-numbers.htm

www.thecrazytourist.com/15-fun-interesting-facts-albania/

www.worldatlas.com/articles/fun-facts-about-albania.html

www.insider.com/jk-rowlings-rejection-letters-2016-7

en.wikipedia.org/wiki/Harry_Potter

www.holtsauto.com/redex/news/3-roads-around-the-world-where-speed-limits-dont-apply/

en.wikipedia.org/wiki/Transport_in_the_Isle_of_Man

www.vox.com/2014/12/11/7373977/fastest-speed-limits

2.3 Linear Systems Everywhere

Halloween around the World

Halloween is very popular around the world. Costumes, trick-or-treating, and jack-o'-lanterns come to mind while thinking of Halloween. Modern-day Halloween came from the ancient Celtic religion celebration of Samhain, meaning *End of Summer* in Celtic. For one night before winter, people carved Jack-o'-lanterns out of turnips to scare evil spirits, dressed in costumes to hide themselves from the souls of dead, and went from house to house telling stories and getting food. The souls of dead ancestors were honored. Samhain parade is still a yearly tradition in Ireland and Scotland. Japanese also have a colorful Halloween parade though there is no trick-or-treating. In Italy, people celebrate both, the modern Halloween and Ognissanti or All Saint's Day when ancestors are remembered. Mexico and Guatemala have the Day of the Dead and a giant kite festival with thousands of brightly colored kites to honor the dead. Kids get candies from their parents on behalf of their dead relatives.

Problem. When Alarico, Blake, Claudia, Dante, and Emma counted candies after Halloween, they were surprised that each pair of kids got a multiple of ten candies. Indeed, together Alarico and Blake had 10 candies, Blake and Claudia got 20 candies, Claudia and Dante gathered 30 candies, Dante, and Emma had 40 candies, Emma and Alarico had 50 candies. How many candies did each kid have?

Answer. No solution

Shopping Cart on Wheels

Advertisements, sales, and other marketing ideas have been created to make shoppers buy items. Market owners and managers have always been thinking of how to make shopping easier and more convenient for customers.

A wonderful idea stroked the owner of the Humpty Dumpty supermarket chain in Oklahoma, Sylvan Goldman. While in his office, Mr. Goldman accidentally put a basket on the seat of a folding chair on wheels and... the first *folding basket carrier*, as he named it, was born. The invention was not immediately picked up. It has been modified over the years. Called a *shopping cart* in American English, a *trolley* in British and Australian English, a *buggy* or a *cart on wheels* is widely used today.

Problem. In what year did the first grocery cart on wheels appear in stores? Use the minimum number of hints below to answer the question and explain why other combinations of hints do not work in order to get a unique solution.

A. Twice the number formed with the first two digits of the year is 1 less than the number formed with its last two digits in the order they appear in the year.
B. The difference between the number formed with the last two digits and the number formed with the first two digits is 1 more than the number formed with the last two digits.
C. The shopping card was introduced in the 2nd millennium.

D. The shopping card was introduced in the 20th century.
E. The shopping card was introduced by Sylvan Goldman.

Answer. 1937

Camel Caravan

A camel train or caravan has carried people and merchandise for centuries if not millennia. People of North Africa and the Arabian Peninsula used the ability of camels to tolerate harsh conditions for their travelling and transporting goods in the desert areas as far as to Nigeria, Cameroon, and Kenya, and even for trading with India, Mongolia, and China.

Problem. Two heavily laden camels were walking along side by side. The first camel complained of the weight of his load. The second camel replied:

- if they gave me three sacks you are carrying, then my load would be three times as heavy as yours.
- if they gave you one of my sacks, our load would be the same.

How many sacks are they carrying if all sacks weigh the same?

Answer. 7, 9

The Tallest Mountains

What mountain is the closest to the stars? Do not rush to say it is Mount Everest. Are you surprised? It is true, Mount Everest, located in Nepal and Tibet, is the highest point on Earth above sea level, but Mount Chimborazo is the Earth's highest point. Its summit is much farther from the Earth's center than the Everest's peak. Why? Because our planet is not round, but rather thicker at the Equator. However, this inactive volcano in the Andes is not higher than the summit of Mount Everest, if measured from sea level. Moreover, despite keeping the title of the highest mountain in Ecuador and being closest to the stars, Mount Chimborazo rates as the 39th highest peak in the Andes.

What is the tallest mountain on Earth? Again, not Everest, but Mauna Kea if measured from its peak to base that is deep beneath the Pacific Ocean, though its peak is 4207.3 m (13,803 ft) above sea level. The volcano Mauna Kea is the highest point in the state of Hawaii, USA.

Problem. The peak of Mount Everest above sea level is 2580 m higher than the peak of Mount Chimborazo. Mauna Kea is 1362 m taller than Everest. The sum of the heights above sea level of both mountains is 4906 m greater than the Mauna Kea's height. How tall are Mount Everest, Mount Chimborazo, and Mauna Kea?

Answer. 8848 m, 6268 m, 10,210 m

Countries with Lakes

Lakes and rivers are the natural wealth of any nation. Water resources are habitats for fish and supply water for different animals. Humans have settled around lakes and rivers since ancient times. The first civilizations were born here. In addition to playing a vital role in cleaning our environment and providing great recreation facilities and beautiful scenery, water resources are used for agricultural, industrial, and household purposes.

Problem. Use the minimum number of hints to find the number of lakes with the size of at least 0.1 km^2 in Canada, Russia, and the USA. Round results to the nearest hundred. Explain why other hints are not needed.

1. Canada, Russia, and the USA have the most numbers of lakes greater than 0.1 km^2.
2. There are approximately 1184 lakes in Canada, Russia, and the USA with an area of at least 0.1 km^2.
3. The Bahamas, Malta, and the Maldives are too small to have lakes.
4. Canada has 75,000 more lakes than four times the number of lakes in Russia.
5. Canada has 59,800 more lakes than eight times the number of lakes in the US.
6. Twice the number of US lakes is 3800 greater that the number of lakes in Russia.

Answer. 879,800, 201,200, 102,500

The Largest Rivers

Running across many nations and countries for thousands of miles, the Amazon River, the Congo River, and the Nile River are the world's leaders in the total drainage basins. These magnificent rivers with unique biodiverse areas have had a spiritual meaning and played a vital role in the development of ancient civilizations, kingdoms, and tribes that settled along their banks. The advanced Egyptian civilization, powerful Kingdom of Kongo, mysterious Inca and Huari civilizations are just a few examples.

The Nile River or, precisely, Nile–White Nile–Kagera–Nyabarongo–Mwogo–Rukarara, passes through Ethiopia, Eritrea, Sudan, Uganda,

Tanzania, Kenya, Rwanda, Burundi, Egypt, Democratic Republic of the Congo, and South Sudan before reaching the Mediterranean Sea. Originating in Peru, the Amazon River brings its waters through Ecuador, Colombia, Venezuela, and Brazil to the Pacific Ocean. It competes with the Nile River for the title of being the longest river in the world. Formed by the waters of the Lualaba and Chambeshi Rivers, the Congo River drains into the Atlantic Ocean after passing Democratic Republic of Congo, Congo, Cameroon, the Central African Republic, Burundi, Tanzania, Zambia, and Angola. Cairo, Manaus Kinshasa, Brazzaville, and other capitals and cities were founded at the banks of these rivers.

Problem. The total drainage basin of the Amazon River is greater than twice the basin area of the Nile by 490 km^2 and is greater than the Congo River basin by 3320 km^2. The basin of the Congo River is greater than the one of the Nile by 425 km^2. The total drainage basins of both Nile and Congo Rivers are 65 km^2 less that the one of the Amazon River. Find the drainage basins of the Amazon, Nile, and Congo Rivers. Are all the conditions of the problem needed?

Answer. 7000 km^2, 3255 km^2, 3680 km^2, no

Alphabets

There are over seven thousand spoken languages around the world. Different symbols have been developed and evolved over millenniums to depict the phonemic structure of words and represent them in writing. Some languages have complex writing systems, others use a set of characters to represent sounds, vowels, or consonants, from which words can be composed. Such sets are called alphabets. The word *alphabet* that links first two letters of the Greek alphabet, *alpha* and *beta*, was first used in its Latin form *alphabetum* in the second century, though it is believed that the first alphabet was originated in the Eastern Mediterranean between 1700 and 1500 BC. Some alphabets have been adjusted to represent other languages by adding or eliminating some letters. Classical Latin was derived from the Etruscan alphabet taking 21 of its symbols and adding new ones. English, French, Italian, and Spanish alphabets are based on Latin scripts, though some letters have been removed or added. All letters in Arabic language are consonants showing vowels by a diacritical mark added to the existing letters, while the Russian alphabet has vowels, consonants, and modifiers.

The alphabet of an indigenous language to Papua New Guinea, called Rotokas, contains the smallest number of letters, as opposed to the Khmer language that has the largest alphabet in the world as recorded by Guinness World Records. Some 16 million citizens of Thailand, Vietnam, Laos, Cambodia, and other Southeast Asian countries speak Khmer.

Problem. Considering the statements below, find the number of letters in Arabic, Classical Latin, Greek, Khmer, Rotokas, and Russian alphabets.

A. The total number of letters in the Greek, Classical Latin, and Russian alphabets is 6 more than in the Khmer alphabet.
B. Twice the number of letters in the Russian alphabet is 4 letters more than the difference in numbers of letters in the Khmer and Rotokas alphabets.
C. The Russian alphabet has 10 letters more than the Classical Latin alphabet.
D. Twice the number of letters in the Khmer alphabet is one more than the sum of triple the number of letters in the Russian alphabet and four times the number of letters in the Rotokas alphabet.
E. Twice the number of letters in the Arabic alphabet is the same as the total number of letters in the Classical Latin and Russian alphabets.
F. The sum of the number of letters in the Russian alphabet and four times the number of letters in the Rotokas alphabet is 7 less than the numbers of letters in Khmer alphabet.

Answer. 28, 23, 24, 74, 12, 33

Notable Minarets

A minaret is a tower with balconies or open galleries built into or adjacent to mosques from which the faithful are called to prayer five times a day. Being symbols of Islamic presence and architectural masterpieces, minarets are constructed in a variety of forms and from different materials.

The tallest minaret in the world is the minaret of the Djamaa el Djazaïr in Algiers, Algeria, while the tallest minaret built of bricks is the Fateh Burj in Chappar Chiri, India. The Fateh Burj or the *Victory Tower* in English is also the tallest tower in India. The Minaret of Jam was built around 1190 entirely of baked bricks and is famous for its intricate brick. The minaret is in a remote and barely accessible region of the Shahrak District. It is the Afghanistan's first cultural heritage site to be listed by UNESCO.

Problem. How tall in feet are the Minaret of Djamaa el Djazaïr, Fateh Burj Minaret, Minaret of Jam if

A. The sum of heights of Fateh Burj Minaret and the Minaret of Jam is 329 feet less than the height of the Minaret of Djamaa el Djazaïr.

B. The sum of heights of the Minaret of Djamaa el Djazaïr and the Fateh Burj Minaret is 133 feet more than five times the height of the Minaret of Jam.

C. The sum of heights of the Minaret of Djamaa el Djazaïr and the Minaret of Jam is 99 feet more than three times the height of the Fateh Burj Minaret?

Answer. 328 ft, 870 ft, 213 ft

The Honey Country

Malta is one of the smallest countries in the world and the smallest country in the European Union. It consists of several tiny and tranquil islands surrounded by crystal-clear blue and green water of the Mediterranean Sea. The name *Malta* derived from the Greek word *meli,* meaning *honey* due to unique honey produced by an endemic subspecies of bees lived on the island. It is interesting that the country has several times more tourists than there are residents due to its stunning scenery, pleasant climate, and safety.

Malta was first inhabited around 5900 BC. The islands were given to the Order of St. John during medieval times, who ruled them as a vassal state of Sicily, protected them from invasions, and supported the arts and improvements. During the last two millennia the Maltese islands have been ruled by the Phoenicians, Carthaginians, Romans, Byzantines, Arabs, Normans, Sicilians, Spanish, Knights Templars, French, and British. Each of them contributed to its rich history forming its unique heritage and architecture. Malta gained independence after 160 years of British rule but remains a part of the British Commonwealth. As in the UK, Malta has the left side-driving and a red telephone booth. Malta has a long Christian legacy. Many of its churches have two clocks, showing different times. It is said that the one on its right shows the correct time for religious locals, while the one on the left shows the wrong time to confuse Satan from disturbing the mass.

Its capital, Valletta, was the first ever planned city in Europe. It is one of the most concentrated historical areas in the world, according to UNESCO.

Problem. Valletta was planned 35 years after Malta was given to the Order of St. John. Both years can be split into two two-digit numbers, like the year 2024 is split into 20 and 24 with 20 as the first number and 24 as the second one. The second number is composed from the year when Malta was given to the Order of St. John is twice the first number of the same year. The ratio between the first and the second numbers composed from the year when Valletta was planned is 3 to 13. The second numbers composed from the year when Valletta was planned is greater than the second number composed from the year when Malta was given to the Order of St. John. Malta became an independent country 434 years after being giving to the Order. When was Malta given to the Order of St. John, when did it plan its capital, and when did it gain independence?

Answer. 1530, 1565, 1964

Giving Flowers

Flowers have been a valued and appreciated tradition since ancient times. It is unknown when this custom arose, but myths and tales of ancient Greece, Rome, China, and Egypt show the meaning and magic of flowers and the way their color, number, and type passes an essential message. White roses reflect purity and innocence, while red roses express love and pink roses show grace, happiness, and gentleness. Flowers have been associated with gods and goddesses, as poppies and roses with Greek Aphrodite or black sunflowers with Russian Baba Yaga.

The tradition of giving flowers flourished during the Middle Ages. During the Victorian era, elegant bouquets of specific flowers expressed feelings and emotions better than any words. Even the way of giving a bouquet was important. A bouquet of red roses symbolized anger if gifted upside down. White roses asked for a new beginning,

Problem. Andre, Bryan, and Carl brought beautiful roses for their dates Anna, Bryana, and Carla. If Andre bought 2 roses more, then Anna would get the average number of roses that Bryan and Carl gave to their dates. If Bryan brought 10 roses less, then Bryana would keep the average number of roses that Andre and Carl presented to their dates. If Carl tripled the number of his roses, then Carla would be given the average number of roses that Anna and Bryana had. How many flowers did each girl get?

Answer. 8, 16, 4

Solutions

Halloween around the World. Let us add all candies that each pair of kids got, $10 + 20 + 30 + 40 + 50 = 150$. Because each kid was counted twice, the total number of candies that kids had was $150 / 2 = 75$. On the other side, two pairs of kids Clodagh and Dante and Emma and Alarico got $30 + 50 = 80$ candies, which is more than all five kids together including them got, which is impossible.

Probably kids were not strong in mathematics when counting candies.

Shopping Cart on Wheels. Let a and b denote two numbers formed with the first and last two digits in the order they appeared in the year. Hints A and B lead to the system of linear equations

$$\begin{cases} 2a = b-1 \\ b-a = a+1 \end{cases} \Rightarrow \begin{cases} 2a = b-1 \\ 2a = b-1 \end{cases},$$

that has infinitely many solutions. Thus, either A or B is needed.

Hint C combined with Hint A or B leads to several solutions. Indeed, in this case a can be 11, 12, 13, …, 19 and the corresponding b is 21, 23, 25, …, 37.

Hint D combined with Hint A or B leads to the unique solution, $a = 19$, $b = 37$ and the year of 1937.

Hint E is descriptive.

The first shopping cart was introduced on June 4, 1937

Camel Caravan. Let x and y be the load weight of the first and second camels. Following the problem, we get the system

$$\begin{cases} 3(x-3) = y+3 \\ x+1 = y-1 \end{cases} \Rightarrow \begin{cases} 3x-12 = y \\ x+2 = y \end{cases} \Rightarrow 3x-12 = x+2 \Rightarrow x = 7, y = 9.$$

The camels carried 7 and 9 sacks.

The Tallest Mountains. Let us denote the altitude of a mountain by its first letters. Following the problem, we can write a system of three linear equations in three variables and perform steps:

$$\begin{cases} e-c = 2580 \\ k-e = 1362 \\ e+c = k+4906 \end{cases} \Leftrightarrow \begin{cases} c = e-2580 \\ k = e+1362 \\ e+c = k+4906 \end{cases} \Leftrightarrow \begin{cases} c = e-2580 \\ k = e+1362 \\ e+(e-2580) = (e+1362)+4906 \end{cases}$$

$$\Leftrightarrow \begin{cases} c = e-2580 \\ k = e+1362 \\ e = 8848 \end{cases} \Leftrightarrow \begin{cases} c = 6268 \\ k = 10210 \\ e = 8848 \end{cases}.$$

Mount Everest's peak is 8,848.86 m (29,032 ft) above sea level, the summit of Chimborazo is 6268 m (20564 ft) above sea level, Mauna Kea from its underwater base to peak is 10,210 m (33,500 ft).

Countries with Lakes. Let the first letter of a country represent the number of its lakes. Hints 1 and 3 are descriptive. Hints 4, 5, and 6 provide quantitative information. Let us use them. Following the problem, we can write the system of three linear equations in three variables

$$\begin{cases} c-4r=75000 \\ c-8u=59800. \\ 2u-r=3800 \end{cases}$$

Subtracting the second equation from the first one, we get $8u-4r=15200$, which after dividing both sides by 4 becomes $2u-r=3800$, which is the third equation. That is, the system has infinitely many solutions: $r = 2u - 3800$, $c = 8u + 59800$, and u is any positive integer because it represents the number of lakes. It means, the three equations are linearly dependent and one of them is not needed.

Let us consider and follow Hint 2:

$$c + r + u = 8u + 59800 + 2u - 3800 + u = 11u + 56000 \approx 1184000 \text{ or } u \approx 102500,$$

if rounded to the nearest hundreds. Then $r \approx 201200$, $c \approx 879\,800$.

Canada has 879,800 lakes, Russia has 201,200 lakes, and the US has 102,500 lakes with the area of at least 0.1 km^2.

The Largest Rivers. Let the first letter of a river represent its total basin in km^2. Then, following the problem, we can write the system of four linear equations in three variables

$$\begin{cases} 2n+490=a \\ c-n=425 \\ n+c+65=a \\ a=c+3320 \end{cases}.$$

Adding the first and second equations, we get the third equation of the system. Thus, one of these equations is a linear combination of the other two and is not needed. Let us remove the first equation. Then, expressing n and a in terms of c, we get

$$\begin{cases} c-n=425 \\ n+c+65=a \\ a=c+3320 \end{cases} \Leftrightarrow \begin{cases} n=c-425 \\ n+c+65=a \\ a=c+3320 \end{cases} \Leftrightarrow \begin{cases} n=c-425 \\ (c-425)+c+65=(c+3320) \\ a=c+3320 \end{cases}$$

$$\Leftrightarrow \begin{cases} n=c-425 \\ c=3680 \\ a=c+3320 \end{cases} \Leftrightarrow \begin{cases} n=3255 \\ c=3680. \\ a=7000 \end{cases}$$

The drainage basins of Amazon, Nile, and Congo Rivers are 7000 km^2, 3255 km^2, and 3680 km^2 correspondingly.

Alphabets. Instead of constructing and solving a system of six equations in six variables, let us find a smaller closed sub-system. Indeed, Statements B, D, and F describe connections among the numbers of the Khmer k, Russian r, and Rotokas p alphabets and lead to the following system:

$$\begin{cases} 2r-4=k-p \quad (I) \\ 3r+4p=2k-1 \; (II) \\ r+4p=k+7 \quad (III) \end{cases}$$

Adding the first and the third equations and subtracting the second equation, we get

$$2r-4+r+4p-3r-4p=k-p+k+7-2k+1,$$

that produces $p = 12$.

Subtracting the first equation from the third one leads to $3p - r = 3$, that with $p = 12$ gives $r = 33$, and then $k = 74$.

Then, using Statement E, we can find that the Classical Latin alphabet has $33 - 11 = 23$ letters. From Statement F it follows that the number of letters in the Arabic alphabet is $(23 + 33)/2 = 28$ letters. Finally, Statement A hints that the number of letters in the Greek alphabet is $74 + 6 - 23 - 33 = 24$.

Arabic has 28 letters in its language, and the Classical Latin alphabet consists of 23 letters. There are 24 letters in the Greek alphabet and 74 in the Khmer alphabet. The Rotokas alphabet contains just 12 letters, while the Russian alphabet has 33.

Notable Minarets. Let d, f, and j denote the heights in feet of the Minaret of Djamaa el Djazaïr, Fateh Burj Minaret, Minaret of Jam, then

$$\begin{cases} f+j=d-329 \\ d+f=5j+133 \\ d+j=3f+99 \end{cases} \Rightarrow \begin{cases} f-d+j=-329 \\ f+d-5j=133 \\ -3f+d+j=99 \end{cases}.$$

The solution of the system can be presented as

$$\begin{bmatrix} 1 & -1 & 1 & -329 \\ 1 & 1 & -5 & 133 \\ -3 & 1 & 1 & 99 \end{bmatrix} \Rightarrow \begin{bmatrix} 1 & -1 & 1 & -329 \\ 0 & -2 & 6 & -462 \\ 0 & -2 & 4 & -888 \end{bmatrix} \Rightarrow \begin{bmatrix} 1 & -1 & 1 & -329 \\ 0 & 1 & -3 & 231 \\ 0 & 0 & 1 & 213 \end{bmatrix} \Rightarrow \begin{bmatrix} 1 & 0 & 0 & 328 \\ 0 & 1 & 0 & 870 \\ 0 & 0 & 1 & 213 \end{bmatrix}.$$

The Djamaa el Djazaïr in Algiers stands at 870 ft or 265 m. Fateh Burj in Chappar Chiri is 328 ft or 100 me tall. Minaret of Jam is 213 ft tall.

The Honey Country. Let a and b be the first and second numbers formed from the year when Malta was given to the Order of St. John and c and d be the first and second numbers formed from the year Valletta was planned. Then the years can be presented as $100a + b$ and $100c + d$ and

$$\begin{cases} (100c+d)-(100a+b)=35 \\ b=2a \\ d=\dfrac{13}{3}c \\ b<d \end{cases}.$$

From the first equation $100c - 100a + d - b = 35$ follows that either

(a) $c = a$ and $d - b = 35$ or
(b) $c = a + 1$ and $100 + d - b = 35$.

The second option violates the inequality $b < d$, so we choose $c = a$.

Then, the second and third equations lead to $b = 2a$ and $d = 13c/3 = 13a/3$, and the first equation can be rewritten as $\dfrac{13}{3}a - 2a = 35$, from which follows $a = 15$. Then $c = 15$, $b = 30$ and $d = 65$.

The islands were given to the Order of St. John in 1530 that planned Valletta in 1965. Malta gained independence from the UK in 1964.

Giving Flowers. Let a, b, and c be the numbers of roses Andre, Bryan, and Carl brought to their dates. Following the problem, we get the system of three linear equations in three variables:

$$\begin{cases} a+2=\dfrac{b+c}{2} \\ b-10=\dfrac{a+c}{2} \\ 3c=\dfrac{a+b}{2} \end{cases} \Rightarrow \begin{cases} 2a+4=b+c \\ 2b-20=a+c \\ 6c=a+b \end{cases}.$$

Adding the first two equations we get $2a + 2b - 16 = a + b + 2c$ or $a + b - 16 = 2c$, that with the third equation leads to $4c = 16$ or $c = 4$. Substituting this value to the first equation produces $2a = b$. Then, the second equation is transformed to $4a - 20 = a + 4$ or $a = 8$ and $b = 16$,

Andrew brought Anna 8 roses, Bryan gave 16 roses to Berta, and Carl gave Carla 4 roses.

Sources

essexflorist.net/tradition-giving-flowers-become-today

brbgonesomewhereepic.com/20-facts-about-malta/

www.unitedlanguagegroup.com/blog/translation/how-halloween-is-celebrated-around-the-world

americanhistory.si.edu/object-project/refrigerators/cart

en.wikipedia.org/wiki/Shopping_cart

oceanservice.noaa.gov/facts/highestpoint

en.wikipedia.org/wiki/Mauna_Kea

en.wikipedia.org/wiki/Chimborazo

en.wikipedia.org/wiki/Camel_train

en.wikipedia.org/wiki/List_of_rivers_by_length

learnodo-newtonic.com/river-congo-facts

www.discoverwalks.com/blog/egypt/top-10-astonishing-facts-about-the-river-nile

en.wikipedia.org/wiki/Amazon_Riverhttps://en.wikipedia.org/wiki/Nile

www.google.com/search?q=congo+river+countries

en.wikipedia.org/wiki/Minaret

en.wikipedia.org/wiki/List_of_tallest_minaretst

en.wikipedia.org/wiki/Minaret_of_Jam

2.4 Nonlinear Systems through Time

The Newest Country

History has witnessed countries, empires, and kingdoms that disappeared, emerged, split, or united. Serbia, Montenegro, East Timor, Eritrea, and Palau are among 17% of countries formed in the 20th century. Sudan is the latest country divided into two. Its name means *the land of the blacks* in Arabic. South Sudanese have Christian and animist beliefs, while North Sudanese are mostly Muslims. Differences in religion, language, and culture led to a 50-year civil war between North and South Sudan. Even a peace agreement signed between them did not prevent further conflicts. Southerners voted to secede from the north and form the world's youngest country, Republic of South Sudan, with the capital in Juba. South Sudan has become a member of the United Nations and the African Union. Despite rich biodiversity, petroleum, gold, and other natural resources, the majority of its population lives in poverty, without access to education, health facilities and drinking water.

Problem. Find the year when South Sudan gained its independence if this year can be split in two two-digit numbers. The first two digits in the order they appear in the year form the first number and the last two digits in the order they appear in the year form the second number, which is less than the first number. The sum of the two numbers is 31, and the sum of their squares is 521.

Answer. 2011

A Sculpture Lake

Beautiful red-water Lake Natron at the border of Tanzania and Kenya is not like any other lake in the world. It is called a *sculpture lake* because its highly alkaline water from volcanoes turns birds and animals into unique stone mummified statues that can be seen in and around the lake. A high concentration of salt, soda, and magnesite sediments in water supports cyanobacteria that damages almost all living organisms that drink it. Surprisingly, flamingoes thrive and breed there. Lake Natron receives only 400 mm of rain a year that often evaporates before reaching the lake surface where the temperature can rise up to 140 °F.

Problem. Find the area of Lake Natron in square miles if it is the product of two numbers. The sum of the numbers is 73, while the sum of their cubes is 300,979. Is it possible to solve the problem without finding the two numbers first?

Answer. 402 sq. miles, yes

National Flag with Stars

Chinese flags have been changed several times over the country's rich history. Its current national flag has been in use since the foundation of the People's Republic of China on October 1, 1949. The red background symbolizes the Chinese Communist Revolution, and yellow stars on the upper left conner emphasize the unity of Chinese people under the leadership of the Chinese Communist Party.

Problem. The National Flag of the People's Republic of China features big and small yellow stars on the red flag background. The fourth power of the number of small stars is 255 greater than the fourth power of the number of big stars. The difference between squares of the numbers of small and big stars is 15. How many small and big stars are on the National Flag of China?

Answer. 4, 1

Time Zones

Each country, city, and even a small place used to set up and have their own time. Growing communications among nations and fast development of transportation during the 19th century required a unified time system. In 1878, a Scottish Canadian engineer and inventor Sir Sandford Fleming developed the system of worldwide standard time zones which is still in use today. Based on the idea that the Earth completes a 360° rotation in 24 hours, he proposed the planet to be divided into 24 time zones. The boundaries of countries have brought some corrections. Indeed, China is the largest country with only one time zone, instead of designated five. The meeting point of borders of Norway/Finland, Norway/Russia, and Russia/Finland has three time zones at the same time. The largest difference of 26 hours between two countries is between the Howland and Line Islands, though they are very close to each other.

It is interesting that there are three different days for a short period of time in United Kingdom, Niue, and Kiribati.

Problem. France, including its overseas territories, has the most time zones followed by Russia and USA. Russia has the most time zones across one contiguous region. The cube of the sum of numbers of time zones in France and Russia is 12,167. The sum of cubes of these numbers is 3059. How many time zones are in France and Russia?

Answer. 12, 11

Smallest Continent, Largest Island

Australia is the oldest, flattest, driest, and smallest continent with the least fertile soils. Australia is the only country that covers the entire continent. It is also the only place in the world for kangaroos, the most venomous snake, the box jellyfish, and the blue ringed octopus.

The largest island Greenland is between the Arctic and Atlantic oceans. Greenland is an autonomous country within the Kingdom of Denmark. One of the legends states that, while in exile, Norse explorer and murderer Erik the Red named the island covered with glaciers, ice, and snow *Greenland* in hope that this would attract people. Anyway, scientists state that Greenland used to be quite green 2.5 million years ago.

Problem. Find the area in mil square miles of Australia and Greenland, if the difference of six powers of these areas is 594.561177, and the difference of their cubes is 23.877.

Answer. 2.9 mil sq. miles, 0.8 mil sq. miles

Countries around the Globe

Each sovereign country is defined by a border over a certain territory. Several sovereign countries can be emerged from one country as from Yugoslavia or Soviet Union, or united to one country, as East and West Germany did. Africa is the second largest and second most populous continent, after Asia. On the other side, Europe is the second smallest continent though the third largest populous one.

Problem. There are more sovereign countries in Africa than in Europe. Together they have 99 countries. The sum of inverses of countries in Europe and Africa is 11/270. How many countries are in Europe and Africa?

Answer. 45, 54

The Narrowest Country

Chile is the world's narrowest and longest country, the southernmost country, and the closest to Antarctica. The Republic of Chile occupies a long, narrow strip of land between the Andes and the Pacific Ocean in the western part of South America. It is believed that its name is related to Native American words meaning *ends of the Earth*, or *sea gulls*, or *the deepest point of the Earth.* Chile borders Peru, Bolivia, and Argentina.

Problem. The narrowest and widest width in kilometers of Chile is hidden in one arithmetic sequence. The narrowest width can be presented in different ways, as the first term of the arithmetic sequence to the sixth power or as its second term cubed, or as its fourth term squared. The widest width of Chile is presented by a three-digit number. The third term of the arithmetic sequence is its ones-digit. The numbers one less and one more than the second term of the sequence are its hundreds and tens digits. What are the smallest and largest width of Chile?

Answer. 64 km, 356 km

The Lowest Points of the Continents

The Dead Sea in Israel and Jordan is the lowest dry point not only in Asia but also on the Earth. It is fed by the Jordan River and has deposits of asphalt which float upon its surface. A crater and volcanic Lake Assal in Djibouti is the African lowest elevation. Both water areas are surrounded by deserts and are highly saline that makes it hard for marine life to survive. The lowest point in South America, Laguna del Carbón in Argentina, is the lowest point in the Southern and Western Hemispheres. The lowest point in North America is in the dry and hot Badwater Basin of Death Valley in the US. Deep Lake of the Vestfold Hills is the lowest point in Antarctica, while Lake Eyre is the Australian lowest point. The Caspian Sea shared by Russia, Azerbaijan, Kazakhstan, Turkmenistan, and Iran is the lowest point in Europe.

Problem. The lowest point on dry land of the Dead Sea is 1000 feet less that the square of the lowest point of Lake Eyre, 25 feet more than four times the elevation below sea level of Laguna del Carbón, 126 feet lower than three times the lowest point of Lake Assal, and just 9 feet lower than five times the lowest point of Death Valley. Deep Lake is 13 feet below three times Lake Eyre, which is 43 feet higher than the Caspian Sea. The difference between squares of elevations below sea level of Laguna del Carbón and Lake Eyre is 115935. Find the lowest points on each continent. All given elevation values are integers.

Answer. –1,401 ft, –509 ft, –344 ft, –282 ft, –160 ft, –92 ft, –49 ft

Quadripoints

Neighboring countries are separated by borders between them. The borders between several countries can meet at one point. The tripoint, where three countries meet, can be in the middle of a river or lake, on mountain summit, or at a city. There are over 157 such tripoints. Finland, Norway, and Sweden meet at the Treriksröset tripoint, Argentina, Brazil, and Paraguay meet at the Triple Frontier, and India, Pakistan, and China meet at the summit of Sia Kangri. Bratislava is the only capital in the world located at a tripoint where Slovakia, Hungary, and Austria meet. Four African countries, Zambia, Zimbabwe, Botswana, and Namibia, come together on one spot, called a quadripoint, at the middle of the Zambezi River. There are no points where five or more countries meet except for the South Pole.

Problem. How many quadripoints in addition to the African quadripoint are in the world if their number is a solution x that satisfies the following system

$$\begin{cases} y = 3x^2 + 2 \\ x = 2y^2 + 3 \end{cases}?$$

Answer. none

A Guinness World Record Holder

The legendary Cuban leader Fidel Alejandro Castro Ruz (1926–2016) led an armed revolution against President Fulgencio Batista and entered the capital city of Havana on 1 January 1959. He has been included to the Guinness Book of Records several times for being the world's third longest-serving head of state after Britain's Queen Elizabeth II and the King of Thailand, for holding the longest continuous 49-year tenure as a national non-royal leader, for giving the longest speech at the United Nations in 1960, and for surviving the most number of assassination attempts that involved poison pills, a toxic cigar, chemically tainted diving suit, and others. *Time Magazine* in 2012 named Fidel Castro as one of the 100 most influential personalities of all time.

Problem. Fidel Castro had survived the greatest number of assassination attempts. Find this number if it can be represented by four numbers

- as the sum of the first number squared and the second number,
- as the difference between the third number squared and the fourth number,
- the third number is one more than the first number,
- the fourth number is the sum of the first two numbers. The sum of the four numbers is 102.

Answer. 638

Solutions

The Newest Country. Let a and b be the first and the second numbers, $a > b$. Then

$$\begin{cases} a+b=31 \\ a^2+b^2=521 \end{cases}.$$

Substitution of $b = 31 - a$ to the second equation leads to the quadratic equation $a^2 - 31a + 220 = 0$, that has a solution

$$a=\frac{31\pm\sqrt{31^2-4\cdot1\cdot220}}{2\cdot1}=\frac{31\pm9}{2},$$

$a = 20$ and $b = 11$ or $a = 11$ and $b = 20$. Since $a > b$, only the first set satisfies the problem.

The world's newest country South Sudan appeared on July 9, 2011.

A Sculpture Lake. Let a and b be two numbers. Then

$$\begin{cases} a+b=73 \\ a^3+b^3=300979 \end{cases}.$$

1st way. We are interested in the product ab. Therefore, instead of solving the system for a and b, let us concentrate on finding the product ab. Dividing the second equation by the first one and remembering the formula $a^3 + b^3 = (a + b)(a^2 - ab + b^2)$, we get $a^2 - ab + b^2 = 4123$. Squaring the first equation leads to $a^2 + 2ab + b^2 = 5329$. Combining the last two equations, we get $ab = 402$.

2nd way. Substitution of $b = 73 - a$ to the second equation leads to the equation $a^3 + (73 - a)^3 = 300979$, that, after using $(a - b)^3 = a^3 - 3a^2b + 3ab^2 - b^3$, becomes $a^3 + (389017 - 15987\,a + 219a^2 - a^3) = 300979$, or the quadratic equation $219a^2 - 15987\,a + 88038 = 0$, that has a solution

$$a = \frac{15987 \pm \sqrt{15987^2 - 4\cdot 219\cdot 88038}}{2\cdot 219},$$

$a = 67$ with $b = 6$ and $a = 6$ with $b = 67$. Their product ab is 402.

The area of Lake Natron is 402 square miles.

National Flag with Stars. Let $s > 0$ and $b > 0$ be the numbers of small and big stars on the Chinese National Flag. Then,

$$\begin{cases} s^4 - b^4 = 255 \\ s^2 - b^2 = 15 \end{cases}.$$

Remembering $s^4 - b^4 = (s^2)^2 - (b^2)^2 = (s^2 - b^2)(s^2 + b^2)$ and using the second equation, the first equation can be simplified to

$$255 = s^4 - b^4 = 15\cdot(s^2 + b^2).$$

Then the system becomes

$$\begin{cases} s^2 + b^2 = 17 \\ s^2 - b^2 = 15 \end{cases} \Rightarrow \begin{cases} s^2 = 16 \\ b^2 = 1 \end{cases} \Rightarrow b = \pm 1, s = \pm 4.$$

Since $s > 0$ and $b > 0$, we take $s = 4$ and $b = 1$.

The Chinese National Flag has one big star surrounding by four small stars for workers, peasants, middle class citizens, and solders.

Time Zones. Let $f > 0$ and $r > 0$ be the number of time zones in France and Russia. Then

$$\begin{cases} (f+r)^3 = 12167 \\ f^3 + r^3 = 3059 \end{cases}.$$

Taking the cube root from both sides of the first equation we get, $f + r = 23$, or $r = 23 - f$, that in combination with the second equation leads to

$$f^3 + (23-f)^3 = 3059 \Rightarrow f^3 + 23^3 - 3 \cdot 23^2 f + 3 \cdot 23 f^2 - f^3 = 3059$$

$$\Rightarrow 3f^2 - 69f + 396 = 0.$$

The solution to the last quadratic equation is

$$f = \frac{69 \pm \sqrt{69^2 - 4 \cdot 3 \cdot 396}}{2 \cdot 3} = \frac{69 \pm 3}{6}.$$

Thus, two solutions are possible, $f = 12, 11$ and the corresponding $r = 11, 12$. Because France has the most number of time zones, $f = 12$, and $r = 11$ are taken.

France has 12 time zones mostly due to its territories around the world. France, as a European country, has just one time zone. Russia has 11 contiguous time zones.

Smallest Continent, Largest Island. Let a and g be the areas of Australia and Greenland in mil square miles. Following the problem, we can write the system of equations

$$\begin{cases} a^6 - g^6 = 594.561177 \\ a^3 - g^3 = 23.877 \end{cases} \Rightarrow \begin{cases} (a^3)^2 - (g^3)^2 = 594.561177 \\ a^3 - g^3 = 23.877 \end{cases}.$$

Using the formula $a^2 - b^2 = (a-b)(a+b)$ for the difference of squares, we obtain

$$\begin{cases} (a^3 - g^3)(a^3 + g^3) = 594.561177 \\ a^3 - g^3 = 23.877 \end{cases}.$$

Substituting the second equation to the first one, we get

$$\begin{cases} a^3 + g^3 = 24.901 \\ a^3 - g^3 = 23.877 \end{cases}.$$

Adding two equations leads to $2a^3 = 48.778 \Rightarrow a^3 = 24.389 \Rightarrow a = 2.9$. Then $g = 0.8$.

The area of Australia is 2.9 mil sq. miles, the area of Greenland is 0.8 mil sq. miles.

Countries around the Globe. Let a and e be the number of countries in Africa and Europe, $a > e$. Following the problem,

$$\begin{cases} a+e=99 \\ \dfrac{1}{a}+\dfrac{1}{e}=\dfrac{11}{270} \end{cases}.$$

The second equation leads to $\dfrac{a+e}{ae}=\dfrac{11}{270}$, that in combination with the first equation becomes $\dfrac{99}{ae}=\dfrac{11}{270}$ or $ae = 2430$. Then

$$\begin{cases} a+e=99 \\ ae=2430 \end{cases}.$$

The system has two solutions $a = 54$, $e = 45$ and $a = 45$, $e = 54$. We choose the first one because $a > e$.

There are 45 countries in Europe and 54 in Africa.

The Narrowest Country. Let a and d be the first term and difference of an arithmetic sequence. Since a represents the width, it is greater than 0, $a > 0$. Moreover, from the problem follows that $d > 0$. Then $(a + d)$ and $(a + 3d)$ are its second and fourth terms and, following the problem, we can write

$$\begin{cases} a^6=(a+d)^3 \\ a^6=(a+3d)^2 \end{cases}.$$

Taking the cube root from both sides of the first equation and the square root from the second one, we get

$$\begin{cases} a^2=a+d \\ a^3=a+3d \end{cases} \Rightarrow \begin{cases} a^2-a=d \\ a^3-a=3d \end{cases}.$$

1st way. The substitution d from the first equation to the second equation leads to

$$a^3-a=3a^2-3a \Rightarrow a^3-3a^2+2a=0 \Rightarrow a(a^2-3a+2)=0.$$

Since $a > 0$, we get the quadratic equation $a^2 - 3a + 2 = 0 \Rightarrow a = 1$ and then $d = 0$ or $a = 2$ with the corresponding $d = 2$. The option $d = 0$ does not satisfy the problem, namely, it does not produce a three-digit number for the largest width. Hence, the lowest width is $2^6 = 64$.

2nd way. Remembering that $a > 0$ and $d > 0$ and dividing the second equation by the first one, we get $a = 2$.

The terms of the arithmetic sequence are 2, 4, 6, 8,… . Therefore, the last digit of the widest width is 6, the first and the second are 3 and 5, and the width is 356 km.

Chile ranges from 64 to 356 km across.

The Lowest Points of the Continents. Let s, f, l, v, a, c, and e in feet be elevations below see level of the lowest points of Dead Sea, Lake Assal, Laguna del Carbón, Death Valley, Deep Lake, Caspian Sea, and Lake Eyre. Following the problem, we can write down the system

$$\begin{cases} e^2 = s + 1000 \\ 4l = s - 25 \\ 3f = s + 126 \\ 5v = s + 9 \\ a = 3e + 13 \\ c = e + 43 \\ l^2 - e^2 = 115935 \end{cases}.$$

A careful look at the system tells us that the first, second, and seventh equations produce a closed system in three equations

$$\begin{cases} e^2 = s + 1000 \\ 4l = s - 25 \\ l^2 - e^2 = 115935 \end{cases}.$$

Combining the first and third equation leads to

$$\begin{cases} 4l = s - 25 \\ l^2 - s - 1000 = 115935 \end{cases}.$$

1st way. Squaring both sides of the first equation and subtracting it from the second equation multiplied by 16 produces the quadratic equation $s^2 - 66s - 1870335 = 0$ that has two solutions

$$s = 33 \pm \sqrt{33^2 + 1870335} = 33 \pm 1368,$$

Namely, $s = 1335$ and $s = 1401$. Knowing s, let us find l from $4l = s - 25$ as $l = 327.5$ for $s = 1335$ and $l = 344$ for $s = 1401$. The first number 327.5 is not integer, then only $s = 1401$ works. Using this value and other equations of the original system of seven equations, we get $e = 49, f = 509, v = 282, a = 160, c = 92$.

2nd way. Substituting $s = 4l + 25$ to the second equation leads to $l^2 - 4l - 116{,}960 = 0$ that has two solutions $l = -340$ and $l = 344$. Only the second solution satisfies the problem and leads to a desirable result.

The Dead Sea sits at an elevation of –1401 feet below sea level. Elevations of Lake Assal is 509 feet, of Laguna del Carbón is 344 feet, of Death Valley is 282 feet, of Deep Lake –160 feet, of Caspian Sea 92 feet, of Lake Eyre 49 feet below sea level.

Quadripoints. Instead of solving the system of quadratic equations, let us sketch their graphs. Since both equations are quadratic in one variable with a positive leading coefficient, their graphs are represented by two parabolas, one is open up, the second one is open to the right (see Figure 2.1).

It is clearly seen that the parabolas do not have any point of intersection. It means the system does not have any solution.

Another way to solve the system is to substitute $x = 2y^2 + 3$from the second equation to the first one and solve the obtained equation of the fourth degree $12y^4 + 36y^2 - y + 29 = 0$.

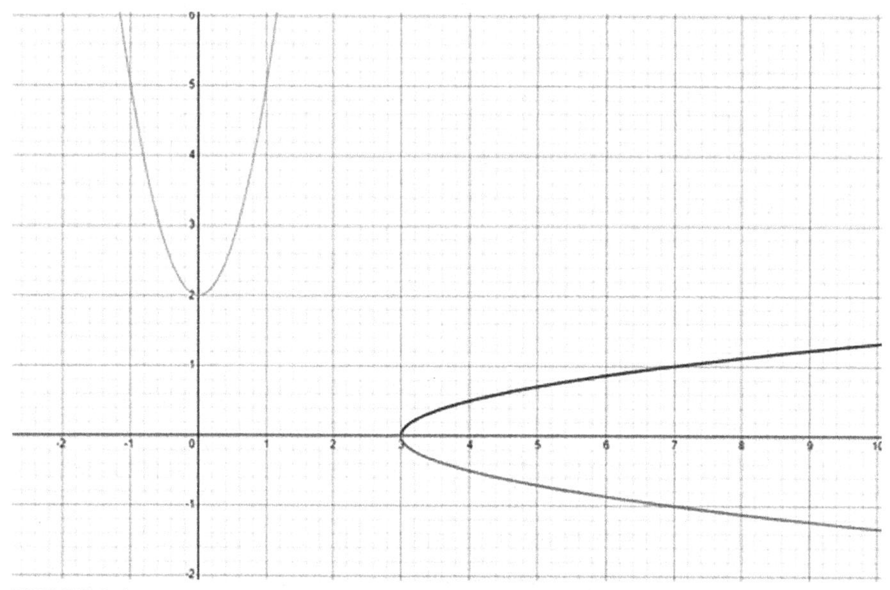

FIGURE 2.1

There is only one quadripoint in the world where the corners of four nations, Zambia, Zimbabwe, Botswana, and Namibia, come together.

A Guinness World Record Holder. Let n and m be the first and the second numbers, then $(n + 1)$ and $(n + m)$ are the third and the fourth numbers and

$$\begin{cases} n^2+m=(n+1)^2-(n+m) \\ n+m+(n+1)+(n+m)=102 \end{cases} \Rightarrow \begin{cases} n^2+m=n^2+2n+1-n-m \\ 3n+2m+1=102 \end{cases} \Rightarrow \begin{cases} n-2m=-1 \\ 3n+2m=101 \end{cases}.$$

Adding the two equations from the last system, we get $n = 25$, and then $m = 13$ and $25^2 + 13 = 638$.

Fidel Castro survived 638 assassination plots.

Sources

facts.net/country-facts/

en.wikipedia.org/wiki/Tripoint

www.atlasobscura.com/places/african-quadripoint

en.wikipedia.org/wiki/Flag_of_China

en.wikipedia.org/wiki/Australia

www.funkidslive.com/learn/top-10-facts/top-10-facts-about-australia/

visitgreenland.com/articles/10-facts-nellie-huang/en.wikipedia.org/wiki/Greenland

followalice.com/knowledge/lake-natron-tanzanias-beautiful-and-deadly-red-lake

www.cheapflights.co.uk/news/the-10-newest-capital-cities

en.wikipedia.org/wiki/South_Sudan

www.britannica.com/place/South-Sudan

borgenproject.org/facts-about-south-sudan/

www.infoplease.com/calendars/history/time-zone-origins

citymonitor.ai/community/here-are-some-worlds-most-stupid-time-zones-2863

www.worldometers.info/geography/how-many-countries-in-africa

www.schengenvisainfo.com/countries-in-europe/

www.guinnessworldrecords.com/world-records/65331-longest-serving-head-of-state-non-royal

everwww.abc.net.au/triplej/programs/hack/how-castro

www.panarmenian.net/eng/news/86837/

https://www.nbcnews.com/storyline/fidel-castros-death/fidel-castro-death-cuban-leader-held-world-record-n688516

en.wikipedia.org/wiki/Chile

www.nationsencyclopedia.com

www.worldatlas.com/articles/the-lowest-points-of-the-world-s-continents.html

3

Wor(l)d Problems

Mathematics compares the most diverse phenomena and discovers the secret analogies that unite them.

~ Joseph Fourier (1768 –1830)[1]

Work, mixture, distance, and other commonly used word problems are the subjects of seven sections of this chapter. You will travel across countries, learning their history and heritage and visiting their capitals and markets. You will be excited to learn about the world's major religions and discover the story of football, dominoes, and other popular games. Some of these problems are a little challenging – their solutions might not be obvious, require comprehensive analysis, or possess unexpected tricks of identifying relevant information, deciding on variables, and combining rules and formulas. Persisting through such twists will equip you with expertise and develop the confidence in you necessary for solving variety of word problems and finding meaningful solutions with real-life interpretations. Besides gaining mathematical proficiency and mastery, you will learn inspiring stories about how not giving up on your dreams can turn them into reality. Get ready for a wor(l)d of fun!

3.1 Profit at Markets

The Vernissage Market

Armenia is a landlocked mountain country in Western Asia. Hayk is the legendary patriarch of Armenians and a great-great-grandson of Noah. He established a new nation in the Ararat region after defeating Babylonian King Bel in 2492 BCE. Yerevan has been the capital of Armenia since 1918 and one of the world's oldest continuously inhabited cities. Various unique places can be visited in the city and country. Among them is the Vernissage market

[1] A famous French mathematician and physicist who developed the Fourier series, Fourier analysis, Fourier transform.

DOI: 10.1201/9781003351702-3

on Sundays. It offers Armenian folk art, unique jewelry, bronze, silver, ceramic, and other exclusive handmade masterpieces. *Vernissage* is French for *vanishing*, meaning a private preview of an art exhibition before it formal opening.

Problem. A tourist bought a silver souvenir the Vernissage market for 10,000 plus half the original price in Armenian drams. What was the original price of the souvenir?

Answer. 20,000

An Eritrean Market

The State of Eritrea in the Horn of Africa was named by Italians in 1890s and reaffirmed by the 1993 independence referendum. *Eritrea* comes from the ancient Greek word meaning *red*. Indeed, the northeastern and eastern parts of the country have a long coastline along the Red Sea, while its land borders Ethiopia, Sudan, and Djibouti. Through its rich history Eritrea was a part of Kingdoms of D'mt, Aksum, and Medri Bahri, the Ottoman empire, and Italy, just to mention a few, before gaining its independence from Ethiopia in 1993. Being a multilingual and multicultural country, Eritrea has no official language.

The capital city of Eritrea, Asmara, also referred to as *New Rome* or *Italy's African City*, has been declared a UNESCO World City Heritage in 2017. Asmara, ancient ruins, and Keren Camel Market are the main tourist destinations of this beautiful country.

Problem. While visiting Keren Camel Market in Eritrea, Isaias bought a camel for 3000 Eritrean nakfa and sold it for 4000 Nkf during his next visit, then bought it back for 5000 Nkf and sold it again for a $6000 Nkf in a few weeks. How much profit did he make?

Answer. 2000 Nkf

A Fish Market

Think and name a Japanese dish. Yes, definitely! It is sushi prepared with rice, seaweed, and a variety of other ingredients. The best sushi and sashimi can be tasted in Japan. Just visit any fish market there. The world's biggest and most famous fish market is Toyosu Fish Market in Tokyo. The market is known for its iconic tuna auctions that start around 5:00 AM. The market floor is white except for a green floor at the tuna auction area because buyers determine the quality and price of the fish by checking its red meat color that stands out best against a green background. After watching the giant fish auction with notable bids from an observation deck, visitors can taste the freshest sashimi, sushi, and seafood in one of numerous restaurants there.

Problem. An auctioneer said to his friend that the amount he got from the tuna auction differed from what he expected by $x\%$ where x is the solution to

$$x \cdot |x| = -81.$$

How much has the retailer gained or lost?

A. The auctioneer gained 9% more than he expected.
B. The auctioneer got 9% less than he expected.
C. The auctioneer gained 81% more than he expected.
D. The auctioneer got 81% less than he expected.
E. The problem does not have a solution.

Express this amount in Japanese yen.

Answer. B, no

An Indian Market

India is famous not only for its rich history and unique architecture but also for its gold, diamonds, and other precious stones. It was the first country to mine and trade diamonds, and it still maintains the largest cutting and polishing diamond industry in the world. Gems and jewelry manufacturing is a vital part of the economy and one of the fastest growing industries in the country. The Indian jewelry market is second only to the US market.

Sarojini Market in Delhi, New Market in Kolkata, Flea Market in Goa, Fashion Street in Mumbai, and other markets, called *bazaars*, are a shopper's paradise offering an astonishing variety of everything from spices and groceries to trendiest clothes, fashionable footwear, fine jewelry, and hand-made souvenirs. These items are offered at reasonable prices, though bargains are very common and appreciated.

Problem. A girl bought a beautiful necklace at the Johari Bazaar in Jaipur. She was so happy with the deal and asked her parents and a sister to guess the price she paid. Mother gave her estimate of the necklace as 1800 Indian rupees, the sister valued it as 1600 Indian rupees, but Father estimated only 1100 rupees. The girl summarized that none of them had guessed the price correctly. Mother was off by 300 rupees; Father was off by 400 rupees, and her sister was off by only 100 rupees. How much did the girl pay for her necklace?

Answer. 1500 rupees

The Tabriz Bazaar

The Tabriz Bazaar is one of the oldest and largest bazaars in the Middle East. It is also the world's largest covered bazaar and is famous for its merchandise, especially for the best Persian rugs in the world. The Tabriz historic bazaar has been a UNESCO World Heritage Site since 2010.

Located on the Silk Road, the city of Tabriz has been a prosperous commercial and cultural center since antiquity and used to be the capital city of the powerful Safavid Kingdom from the 13th to 16th century.

Problem. A shopper wanted to buy a Persian rug at the Tabriz Bazaar and bargained on the price of 33,000,000 Iranian rials. A merchant agreed to reduce the price down to 20,000,000 IRR. The shopper continued to bargain further, but the merchant said that he had already reduced the price by 40%. Was the merchant correct?

Answer. No

A Floating Market

Fruits, vegetables, fish, and other local produce and goods are sold from boats at floating markets in Myanmar, Thailand, Indonesia, Vietnam, Sri Lanka, Bangladesh, and India. Indeed, the areas adjacent to rivers were the first to be inhabited. Originating in places where water transport played an important role and served as a means of transportation and trade, floating markets have survived centuries bringing history and culture of the region to our days. They have become one of the major tourist attractions giving an opportunity to its guests to experience riverside shopping and enjoy the surrounding fascinating landscape.

One of notable floating markets, the Muara Kuin Floating Market in Indonesia, called *Pasar Terapung*, is located at the junction of Barito and Kuin Rivers. Its history goes back to 1526 when Sultan Suriansyah built a kingdom that later became a city of Banjarmasin nicknamed *Kota Air* or *Water City* for it lies below sea level. Barter or *bapanduk* is a lasting tradition here. Most of the traders are women that not only sell but also cook for you.

Problem. Diah spent 100,000 Indonesian rupiah on a souvenir and marked it up 25% at the Muara Kuin Floating Market. What bargain in % could she agree on in order not to lose money?

Answer. 20%

A Perfume Market

The Union of the Comoros is an archipelago in the Mozambique channel of the Indian Ocean. The country consists of four main volcanic islands and numerous smaller islands. These islands are often referred to as *perfumed islands* because their plants are so fragrant that they have become the main ingredients for making perfumes. The four main islands are represented with four white stars on the national flag. The name of the country comes from the Arabic word for the *moon*. The moon is also depicted on the green triangle of the Comoros flag with yellow, white, red, and blue stripes. Most Comorians are Sunni-Muslims. The Comoros got its independence from France in 1975.

Problem. A company bought bulk of parfum oil at a Comoros market for 100,000 Comorian Francs. When selling it, the company marketed it up 25%. However, because of an unexpected price fluctuation, the company had to offer a discount of 25% to sell the oil. How much did the company make or lose on the parfum?

Answer. –75,000 KMF

A Historic Market

Cairo is the capital of Egypt as well as the largest urban agglomeration in Africa, the Arab world, and the Middle East. One of the most famous and exclusive markets in the world, the Khan Al-Khalili Bazaar, is in the historic center of Cairo in a building that was constructed during the 14th to 15th centuries and was named after its founder the Mamluk prince and well-known merchant Jerksy al Khalili. The Bazaar offers heritage crafts, local products, perfume, spices, coffee, and much more.

Problem. A tourist went to the Khan Al-Khalili Bazaar to buy some oil. He had a choice of getting oil at the same price in any of three containers of spherical, conical, and cylindrical shapes. All containers are of the same height. What radius of the base should conical and cylindrical containers have to have the same amount of oil as a spherical container?

Answer. $\sqrt{2}r, \sqrt{\frac{2}{3}}r$

A Train Market

The Maeklong Railway Market is one of the most unusual markets in the world. It is just 60 km away from the Thailand's capital Bangkok or *Krung Thep*

Maha Nakhon in Thai. The market stretches along a functioning train track. Its stalls are full of fruits, veggies, meats, seafood, snacks, clothing, lotions, and flowers. Unfortunately, all items and products should be removed when a train randomly passes through this market several times a day.

Problem. A tourist found a lotion made of local herbs at the Maeklong Railway Market. He was offered lotion in three containers of spherical, conical, and cylindrical shapes for the same price. The widest widths of all containers were the same. A conical container was twice as tall as a spherical one and three times taller than a cylindrical container. What container had the maximum amount of lotion?

A. A spherical container.
B. A conical container.
C. A cylindrical container.
D. Any of them.
E. Specific information about containers is needed.
F. More information is needed.

Answer. D

A Night Market

Asian night markets offer everything from food prepared following mouth-watering recipes and rich decorations to shopping and live entertainment that reflect diverse cultural melting pots and time-honored traditions blending old and new, foreign and local. These casual open-air bazaars are filled with buzzing sounds at night. Their history goes back to medieval China when rules on nighttime activity were smoothed during the Tang dynasty. Night markets are called *pasar malam* in Malaysia, Indonesia, and Singapore.

Thailand is known for having the world's loudest and showiest night bazaar. Vietnam's night markets are famous for their bright luminosity, glowing lights, and lanterns.

Problem. Visiting an Asian night market, two friends wanted to have traditional ice cream to cool down. One of them preferred ice cream in a waffle cone and another one in a cylindrical carton. Both friends wanted to have the same amount of this delicious treat. Each portion of ice cream was offered with a round (hemisphere) top of ice cream. The height of the cone was b

times its radius. How tall should the carton be if the radius of the base is a times the radius of the base of the cone? Does any random range of values of a lead to a solution?

Answer. $\frac{b+2-2a^3}{3a^2}$, no

Solutions

The Vernissage Market. 1st way. The problem can be easily solved without any mathematics if it is carefully read.

2nd way. A mathematical solution. Let x Armenian drams be the original price of the souvenir, then $x = 0.5x + 10{,}000 \Rightarrow x = 20{,}000$.

The tourist bought a silver souvenir for 20,000 Armenian drams.

An Eritrean Market. Isaias lost 1000 Nkf only after the first transaction, then $6000 - 3000 - 1000 = 2000$.

Isaias made 2000 Nkf profit.

A Fish Market. Since $|-x| = |x| \geq 0 \text{and} x \cdot |x| = -81$, then $x \leq 0$. Hence, $x = -9$. It is not possible to answer the second question because the estimated amount in Japanese yen is not given.

An auctioneer got 9% less than he expected.

An Indian Market. Let x be the necklace price. With her estimate of 1800 Indian rupees, the mother was out by 300 rupees, that is, $|x-1800| = 300$, that splits into two equations $x - 1800 = 300$ or $x - 1800 = -300$ and leads to two solutions $x = 2100$ or $x = 1500$.

Analogously, the father's estimate was $|x-1100| = 400 \Rightarrow x - 1100 = 400$ or $x - 1100 = -400 \Rightarrow x = 1500$ or $x = 700$, and the sister's guess gave $|x-1600| = 100 \Rightarrow x - 1600 = 100$ or $x - 1600 = -100 \Rightarrow x = 1500$ or $x = 1700$. The value $x = 1500$ satisfies all equations.

The girl paid 1500 Indian rupees for her necklace.

The Tabriz Bazaar. There are several ways of solving the problem.

1st way. To find what percentage of 33,000,000 IRR 20,000,000 IRR is, let us divide 20,000,000 by 33,000,000 to get .61 or 61%. Thus, the discount of 39% was less than 40%.

2nd way. To find 40% of 33,000,000 IRR, we multiply 0.4 by 33,000,000 to get 19800000, that is less than 20,000,000 IRR.

It is better for the shopper to choose 40% discount.

A Floating Market. Diah spent 100000 Indonesian rupiah on a souvenir, that with 25% profit is 125000 IDR. Then,

$$\begin{array}{lcl} 125000 \text{ IDR} & - & 100\% \\ 100000 \text{ IDR} & - & x\% \end{array} \Rightarrow x = \frac{100000 \cdot 100}{125000} = 80\%.$$

Diah could agree on at most a 20% discount.

A Perfume Market. A company bought bulk of parfum oil for 100000 KMF and marketed it 25 % higher, that is, for 125000 KMF. Then the company gave a discount of 25%, while selling the parfum. It means, the parfum was sold for the price of 92,500 KMF because

$$\begin{array}{lcl} 125000 \text{ KMF} & - & 100\% \\ x \text{ KMF} & - & 75\% \end{array} \Rightarrow x = \frac{125000 \cdot 75}{100} = 92500 \text{ KMF}.$$

Therefore, the company lost 75,000 KMF.

A Historic Market. Since the heights of all containers are the same, the height of each container can be compared to the height of a sphere $d = 2r$, that is, the volumes of containers

$$V_{sph} = \frac{4}{3}\pi r^3, V_{cone} = \frac{1}{3}\pi r_{cone}^2 h_{cone}, \text{and } V_{cyl} = \pi r_{cyl}^2 h_{cyl},$$

can be modified as

$$V_{sph} = \frac{4}{3}\pi r^3, V_{cone} = \frac{1}{3}\pi r_{cone}^2 \cdot 2r, \text{and } V_{cyl} = \pi r_{cyl}^2 \cdot 2r.$$

Equating all volumes, we get

$$r_{cone} = \sqrt{2}r \text{ and } r_{cyl} = \sqrt{\frac{2}{3}}r.$$

The radius of a circular base should be $\sqrt{2}r$ for a conical container and $\sqrt{\frac{2}{3}}r$ for a cylindrical container.

A Train Market. Since the widest widths of all containers are the same and a conical container is three times taller than a cylindrical container and twice as tall as a spherical container, the volumes of containers

$$V_{sph} = \frac{4}{3}\pi r^3, V_{cone} = \frac{1}{3}\pi r_{cone}^2 h_{cone}, \text{and } V_{cyl} = \pi r_{cyl}^2 h_{cyl},$$

can be rewritten in terms of the radius r of a sphere as

$$V_{sph} = \frac{4}{3}\pi r^3, V_{cone} = \frac{1}{3}\pi r^2 \cdot 2 \cdot 2r, \text{and } V_{cyl} = \pi r^2 \cdot \frac{2}{3} \cdot 2r,$$

because the diameter $d = 2r$ shows how tall a sphere is. Comparing the last formulas, we can conclude that all volumes are the same.

A Night Market. Let r, R, h, and H define the radius of the base and the height of a cone and a cylindrical carton. The volumes of a waffle cone and a cylindrical carton with a hemisphere top are

$$V_{cone} = \frac{1}{3}\pi r^2 h + \frac{1}{2} \cdot \frac{4}{3}\pi r^3 = \frac{1}{3}\pi r^2 br + \frac{1}{2} \cdot \frac{4}{3}\pi r^3 \text{ and}$$

$$V_{cartoon} = \pi R^2 H + \frac{1}{2} \cdot \frac{4}{3}\pi R^3 = \pi (ar)^2 H + \frac{1}{2} \cdot \frac{4}{3}\pi (ar)^3$$

should be equal, that is,

$$\frac{1}{3}\pi br^3 + \frac{2}{3}\pi r^3 = \pi a^2 r^2 H + \frac{2}{3}\pi a^3 r^3,$$

from which follows

$$H = \frac{b + 2 - 2a^3}{3a^2} r.$$

The last equation has applied meaning if $a < \sqrt[3]{\frac{b+2}{2}}$. If $a \geq \sqrt[3]{\frac{b+2}{2}}$, then the problem does not have any solution because the height of the carton cannot be zero or negative.

Sources

en.wikipedia.org/wiki/Armenia
www.indonesia-tourism.com/south-kalimantan/muara_kuin.html
www.shopback.my/blog/world-unusual-markets
en.wikipedia.org/wiki/Floating_market
www.kids-world-travel-guide.com/comoros-facts.html

www.allstarflags.com/facts/worlds-ten-most-colorful-flags/
en.wikipedia.org/wiki/Eritrea
www.worldatlas.com/articles/top-10-interesting-facts-about-eritrea.html
www.rough-polished.com/en/expertise/46501.html
www.revv.co.in/blogs/best-shopping-markets-in-india/
www.globotreks.com/destinations/india/fun-interesting-facts-india/
planetofhotels.com/guide/en/blog/best-markets-in-the-world,
tokyocheapo.com/entertainment/sightseeing/toyosu-fish-market-what-to-know/
www.shopback.my/blog/world-unusual-markets
en.wikipedia.org/wiki/Khan_el-Khalili
en.wikipedia.org/wiki/Bazaar_of_Tabriz
whc.unesco.org/en/list/1346/
www.thrillist.com/travel/nation/best-night-markets-around-the-world

3.2 Work on Work

Air Conditioner

People have long used fire for cooking and warming their homes, while the idea of a cooling system is comparatively new. The necessity of cooling down increased with medical and industrial needs. Tremendous efforts were requested from many inventors before modern *air conditioning* (AC) came to life. Scotland Professor William Cullen suggested evaporating liquids in a vacuum in 1748 that led to creating refrigeration technology. American doctor John Gorrie proposed a small steam engine to cool air so that his patients could stand the Florida heat. Following a request from Brooklyn's Sackett-Wilhelms Lithographic and Publishing Company, a 25-year-old American engineer, Willis Carrier, created a primitive cooling system to reduce humidity around the printer. He also introduced the first practical centrifugal refrigeration compressor in 1922 and later improved it. Carrier's invention moved from US factories and entertainment centers to private residences in the mid-20th century long before it did in other countries boosting fast development of US regions with a hot climate. Today, almost 75% of US homes have AC.

Problem. Several heating, ventilation, and air conditioning (HVAC) companies are ready to install and fix an AC system. Mr. Cool HVAC promises to perform 5/6 of a certain AC repair in 10 hours, Fast HVAC estimates that 6 hours are needed to 3/5 of AC fixing, while Eco HYAC needs only 7.5 hours to have 3/4 of repair done. Which company offers the fastest AC repair?

A. Mr. Cool HVAC
B. Fast HVAC
C. Eco HYAC
D. Mr. Cool HVAC or Fast HVAC
E. Mr. Cool HVAC or Eco HYAC
F. Fast HVAC or Eco HVAC

Answer. F

Reclaiming Land

With growing population and shortage of land, countries have been reclaiming land from lakes, rivers, and oceans for centuries. The Netherlands, Bangladesh, Singapore, and other densely populated countries are in tremendous need of additional space. With 4600 sq. miles of land added, China leads the world in land reclamation from the Yellow Sea and the Yangzi River. Then Netherlands with 2700 sq. miles is followed by South Korea with 600 sq. miles of reclaimed land. Retrieval land has been used for habitation and offered more economic, industrial, and agricultural opportunities. Hong Kong International Airport, the Odaiba island in Japan, and the South Korean business district are just a few examples of developments on land acquired from water.

Problem. The Kingdom of Bahrain has increased its area almost twice by reclaiming its coastline from the Persian Gulf and the Gulf of Bahrain. An emergency to repair a draining system appeared during a reclaiming land project. Two firms promised to repair the draining system in 7 hours and 28 minutes if they worked together, while the first firm needed 14 hours for this repair if it worked alone. How much time did the second company need for the repair if it worked alone?

Answer. 16

The Sagrada Familia

The Sagrada Familia is among the most iconic and ambitious religious temples in the world and the most recognizable symbol in Barcelona. *Sagrada Familia* or *Holy Family* in Spanish is devoted to Jesus, Mary, and Joseph. The construction of this last masterpiece of Antoni Gaudí i Cornet

(1852–1926) began in 1882. Unfortunately, thinking of his project and being careless, Gaudí was hit by a tram and was put into a hospital for the homeless due to his poor clothing. It was too late when his friends found him. Gaudí was buried in the crypt at the Sagrada Familia. His death significantly delayed the completion of the construction in addition to lack of donations and the Spanish Civil War.

The Sagrada Familia was designed so that it could be seen from any point in Barcelona and, at the same time, to offer the best views of the entire city from its top. Its colored glass windows will represent the earth's four elements. Gaudí was inspired by nature and used mathematics and geometric forms while designing the Sagrada Familia. Being a proponent of Catalan culture Antoni Cornet also incorporated elements of Catalan heritage. Despite the church still being under conduction, the Sagrada Familia was listed as a UNESCO World Heritage Site in 2005.

It is interesting, that the Sagrada Familia will be three feet shorter than Barcelona's highest point, Montjuic Hill, because Gaudí believed that no man-made structure should be higher than anything in nature created by God.

Problem. The first artist could decorate the windows in one room in the Sagrada Familia in 4 days and the second one in 6 days. To speed up the work, two artists worked together. In a few days the first artist was assigned to a different project and the second one finished the job. In how many days the windows were decorated if all days were full days (presented by integers)?

Answer. 7

A Beer Bath

Beer is not only for drinking but also for bathing. Yeast, hops, vitamin B, active enzymes, carbonic acid, and other natural ingredients in beer open pores, clean skin, soften hair, ease muscle tension, improve blood circulation, detox the body, and contribute to overall strength. Knowing these cure miracles, people have been bathing in beer since ancient times. Luxury wooden barrels filled with beer can be found in Czech Republic, Iceland, Austria, Japan, Hungary, the USA, and other countries around the world.

Problem. Two hoses are used to fill a wooden barrel with beer. One hose can fill in the barrel within 7 minutes if only it is open. The second hose can do this in 8 minutes. How much time in minutes and seconds is needed to fill in a wooden barrel with beer if both hoses are open?

A. 1 min

B. 2 min 30 sec

C. 3 min 44 sec

D. 3 min 45 sec

E. 7 min 30 sec

F. 15 min

Answer. C

A Modern Medieval Castle

The re-construction of a 13th century Guedelon Castle near Treigny in France currently goes back in time to medieval period. Only hammers to break stones and forge iron, wooden wheels to hoist materials up and down, quarry for stone and sand are used instead of modern tools and technologies. All materials, including wood and stone, are obtained locally. The idea of building the castle began in 1997 after an archaeological survey revealed a medieval fortress near Chateau de Saint-Fargeau. Hundreds of people and volunteers in medieval dresses participate in this unique experimental project each year in hope to restore ancient heritage and better understand the medieval and sustainable construction.

Problem. Two stonemasons are assigned to build one room of the Guedelon Castle. The first stonemason can build 4/5 of the room in 8 days, while 10 days are needed for building 3/7 of the room by the second stonemason. How much time is needed to complete 6/7 of the job if two stonemasons work together?

Answer. 6 days

Recovery after Hurricanes

The most mountainous, most populous, and third largest by area Caribbean country, the Republic of Haiti, shares the island of Hispaniola with the Dominican Republic. *Haiti* means *the land of high mountains* in the indigenous Taíno language. Haiti was re-discovered by Christopher Columbus in 1492 who named it *La Isla Espanola* or *the Spanish Isle*. Haiti became the first independent Latin American and Caribbean nation in 1804, the first country in the Americas to abolish slavery, and the only state in history established by a successful slave revolt. French and Creole are official languages of Haiti, and Voodoo is its officially recognized religion. Its currency is named after a plant Gourd important to Haitians. The Citadel Henry is the largest mountaintop fortress in the Western Hemisphere. The rich cultural heritage of Haiti is reflected in art and music and its wonderful beaches attract thousands of tourists. Sadly, Matthew in 2016, Flora in 1963, Ike in 2008, Cleo in 1964, and other powerful hurricanes have brought a lot of destruction to the country killing hundreds of people.

It is interesting that only one spouse needs to be present for a divorce in Haiti.

Problem. Two companies were hired to repair a neighborhood in Haiti after a destructive Hurricane. The companies estimated that it would take them 12 weeks to complete the repair if they worked together. Due to other unexpected assignments, the companies could not start repair at the same time. The second company worked during the time needed for the first company to complete one third of the repairs if it worked alone. Then the second company left the project. Then the first company worked alone during the time the second company required to complete one eighth of the job. As a result, only 7/12 of the repair was done. How much time was needed for each company to do the repair alone?

Answer. 15 and 60 weeks or 30 and 20 weeks

Notable Tunnels

Tunnels have become an enormous aid in planning highways and railroads. They have shortened distances by going through mountains and under rivers and hidden urban traffic underground. All countries value and maintain these incredible engineering constructions.

San Gotardo tunnel in Switzerland is a part of the A2 motorway between Basel and Chiasso. When this 16.9-km tunnel was finished in 1980 it became the world's longest tunnel. Regrettably, there were almost 900 traffic accidents in the first 15 years. The 17.5-kilometer Jinpingshan tunnel in China leads to the world's highest Jinping dam but is open only to authorized vehicles. This tunnel is also the second deepest tunnel in the world.

Problem. To provide a scheduled maintenance of the San Gotardo, the management company considers bids from three firms. The first and the second firms promise to team up on maintenance and finish it in 30 weeks. The first and the third firms require 48 weeks to provide the same job together. Working together, the second and the third firms can celebrate maintenance completion in 45 weeks. How many weeks does each firm need to complete the job if it works alone? How many weeks to complete the maintenance are needed if the three firms work together?

Answer. $62\frac{14}{23}$ weeks, $57\frac{3}{5}$ weeks, $205\frac{5}{7}$ weeks, $26\frac{2}{11}$ weeks

Remarkable Tunnels

The first modern tunnel built under the Thames in England in 1843 was a huge celebration and a giant step toward the current system of transportation. Tunneling has continued to advance taking a very significant place. At 30 m under the ground, the 18.2-km Yamate tunnel in Japan is not only the world's longest urban tunnel but also the world's second longest tunnel following the 24.5-km Laerdal tunnel in Norway. Some portions of the third longest Zhongnanshan tunnel in Southern China are 1640 meters underground. Unbelievable!

Problem. To provide a scheduled repair of the Laerdal tunnel in Norway, the hiring company considers bids from three firms. The first and the second firms promise to team up and finish repairs in 30 weeks. The first and the third firms require 48 weeks to provide the same job. Working together the second and the third firms can complete repairs in 45 weeks.

There is a weekly payment for the repair job. It means even if a company works for one day a week, it will be paid for one full week. Therefore, the hiring company has decided to hire these the three firms for a certain integer number of weeks. The three firms will start together but then leave when their contract (integer number of weeks) is over. What number of weeks should be assigned to each firm to minimize the cost? Is it important to know the productivity of each company?

Answer. 26, 25, 31, no

The Blue Lagoon

The Blue Lagoon is a geothermal spa in a lava field in Iceland. Its healing water is rich in salts, algae, and other minerals provided by the underground geological layers. A high silica content gives it a milky blue shade. The water temperature in the bathing and swimming area of the lagoon is around 99–102 °F. The water renews every two days. The lagoon is man-made and is a byproduct from a geothermal power plant that generates electricity. This subterranean spa is one of the 25 modern wonders of the world and a popular tourist attraction.

Problem. It takes 12 hours to fill in a spa at the Blue Lagoon if two sources are open. Once the sources were not open at the same time. The first source was open to fill the spa during the time needed for the second source to fill two thirds of the spa. Then the first source was closed, and the second source was open during the time the first source could fill the 3/11 of the spa. As a result, only 3/22 of the spa was left to be filled in. How much time is needed for each source to fill the spa if only it is open? Consider only integer solutions.

Answer. 28 h, 21 h

Construction Projects

Our ancestors pushed construction limits by building impressive colossal structures that still puzzle modern society with questions of how they could be built using tools, technologies, and resources available at that time. The Great Wall of China, Chichen Itza in Mexico, Great Pyramid of Giza in Egypt, and many other magnificent cities, tombs and monuments are mysterious and attractive. Humans cannot stop thinking of creating gigantic new canals, airports, subways, residential, entertainment, industrial complexes, and other extraordinary construction projects.

Dubai's current 21 square mile Al Maktoum International Airport in the United Arab Emirates is still under expansion. It is the largest construction project with no definite completion date. The emerging futuristic metropolis King Abdullah Economic City in Saudi Arabia focuses on education, business, and residential properties. South-North Water Transfer Project in China aims to construct three huge canals that will supply water to the north from three largest rivers. A construction project for a high-speed electric train system in the USA is expected to connect the largest cities in California.

Problem. To provide regular maintenance at Al Maktoum International Airport in Dubai the management intends to hire one company but considers bids from three companies. The third company states that it needs twice as many days as the first company, but charges 20,000 Arab Emirates Dirham AED per day less than the first company, while the second company claims that it takes it just two days more that the first company requests, but its charge is twice less compared to the first company. Moreover, the second and third companies require 13/7 times more time for their joint work on the maintenance than if all three companies work on the maintenance together. What one company should be hired to save funding?

Answer. The decision depends on the amount charged by the first company

Solutions

Air Conditioner. Mr. Cool HVAC can perform 5/6 of repairs in 10 hours. It means, its productivity is

$$\frac{5}{6}\cdot\frac{1}{10}=\frac{1}{12}$$

and the company needs 12 hours to fix the AC.

Another way to come to this outcome is to consider the proportion

$$\begin{array}{ccc} \frac{5}{6} & - & 10 \\ & \searrow\!\!\!\nwarrow & \\ 1 & - & t \end{array} \Rightarrow t = \frac{10\cdot 6}{5} = 12 \text{ hours},$$

where 1 stands for the repair completion. Analogously, the productivity of Eco HVAC is

$$\frac{3}{4}\cdot\frac{2}{15} = \frac{1}{10}$$

and the company needs 10 hours to fix the AC. Finally, the productivity of Fast HVAC

$$\frac{3}{5}\cdot\frac{1}{6} = \frac{1}{10}$$

is the same as of the Fast HVAC.

The Fast HVAC and Eco HVAC work with the same productivity and will fix the AC in 10 hours.

Reclaiming Land. 7 hours and 28 minutes is 448 min. Let 1 denote the completed repair project and x hours or 60x minutes be the estimated time the second firm needed to complete the repair if it worked alone. The first firm requested 14 hours or 14·60 minutes for the repair if it worked alone. Then the productivities of two firms per minute were 1/820 and 1/(60x), and 1/820 + 1/(60x) if they worked together. Hence

$$448\cdot\left(\frac{1}{14\cdot 60}+\frac{1}{60x}\right) = 1 \Rightarrow 448\cdot\frac{x+14}{14\cdot 60x} = 1 \Rightarrow 392x = 6272 \Rightarrow x = 16.$$

The second company completes the repair in 16 hours if it worked alone.

The Sagrada Familia. Let 1 denote the completion of decorating windows, x days be the time two artists worked together, and y days be the time the second artists worked alone. The productivities of artists are 1/4 and 1/6, Then

$$\frac{1}{6}y + x\cdot(\frac{1}{4}+\frac{1}{6}) = 1 \Rightarrow \frac{5}{12}x = 1-\frac{1}{6}y, \text{ then } 5x+2y = 12.$$

Different integers can be tested for x and y, or some initial analysis can be provided first.

Indeed, rewriting the equation $5x+2y=12$ as $5x = 2(6 - y)$, we can conclude that x should be even but not greater than 3. The only choice is $x = 2$, that leads to $y = 1$.

Another way is to see that 5 and 2 are co-primes. Then x should be 2 and $6 - y$ should be 5.

The windows were decorated in 7 days.

A Beer Bath. Let 1 denote filling in a wooden barrel with beer and x minutes be the time needed to do this if two hoses are open. Since the first hose can fill in the barrel in 7 minutes, then its productivity per minute is 1/7, and the productivity per minute of the second hose is 1/8. Hence

$$x\cdot\left(\frac{1}{7}+\frac{1}{8}\right)=1\Rightarrow\frac{15}{56}x=1\Rightarrow x=\frac{56}{15}\text{ minutes or }3\frac{11}{15}\text{ minutes, or}$$

$$\begin{array}{ccc} \frac{11}{15} & - & t \\ & \searrow & \\ 1 & - & 60 \end{array} \qquad t=\frac{11\cdot 60}{15}=44.$$

3 minutes and 44 seconds are needed for two hoses to fill in a wooden barrel with beer.

A Modern Medieval Castle. Let 1 denote the entire job, and x days be the time needed for two stonemasons to build the room. The first stonemason can build 4/5/8 of the room in one day, the second can build 3/7/10 for one day. Hence,

$$x\cdot\left(\frac{4}{5}\cdot\frac{1}{8}+\frac{3}{7}\cdot\frac{1}{10}\right)=\frac{6}{7}\Rightarrow x\cdot\frac{1}{7}=\frac{6}{7}\Rightarrow x=6.$$

Two stonemasons could build one room of the Guedelon Castle in 6 days.

Recovery after Hurricanes. Let x and y be the productivity of the first and second companies per week and 1 be the entire repair job. Then the repair is completed in 12 weeks if two companies work together, that is, $12(x + y)= 1$, and $1/x$ and $1/y$ are the numbers of weeks needed for each company to complete the repair alone. The number of weeks required for the first company to do 1/3 of renovation if it works alone is $1/3 \cdot 1/x$, and $1/3 \cdot 1/x \cdot y$ is the repair done by the second company during that time. The number of weeks necessary for the second company to do 1/8 of repair if it works alone is

$1/8 \cdot 1/y$, and $1/3 \cdot 1/y \cdot x$ is the portion of the repair by the first company during that time. Hence,

$$\begin{cases} 12(x+y)=1 \\ \dfrac{1}{3x}y+\dfrac{1}{8y}x=\dfrac{7}{12} \end{cases} \Rightarrow \begin{cases} y=\dfrac{1}{12}-x \\ \dfrac{\dfrac{1}{12}-x}{3x}+\dfrac{x}{8\left(\dfrac{1}{12}-x\right)}=\dfrac{7}{12} \end{cases}$$

$$\Rightarrow \begin{cases} y=\dfrac{1}{12}-x \\ 450x^2-45x+1=0 \end{cases} \Rightarrow \begin{cases} x=\dfrac{1}{30},\ y=\dfrac{1}{20} \\ x=\dfrac{1}{15},\ y=\dfrac{1}{60} \end{cases}.$$

The companies finish the repair in either 15 and 60 weeks or 30 and 20 weeks if they work alone.

Notable Tunnels. Let 1 denote the entire job and f, s, and t be the number of weeks required by the first, second, and third firm to complete the maintenance if a firm works alone. Then $1/f$, $1/s$, and $1/t$ are the productivities of each firm, and, considering the number of weeks they need if they work in pairs, we get the system

$$\begin{cases} \dfrac{30}{f}+\dfrac{30}{s}=1 \\ \dfrac{48}{f}+\dfrac{48}{t}=1. \\ \dfrac{45}{s}+\dfrac{45}{t}=1 \end{cases}$$

Let us reduce the order of the system by eliminating one equation and one variable, i.e., f. Multiplying both sides of the second equation by 5 and subtracting the resulting equation from the first equation multiplied by 8, we get

$$\begin{cases} \dfrac{240}{f}+\dfrac{240}{s}=8 \\ \dfrac{240}{f}+\dfrac{240}{t}=5 \end{cases} \Rightarrow \dfrac{240}{s}-\dfrac{240}{t}=3.$$

Then the system of two equations in two variables

$$\begin{cases} \dfrac{45}{s}+\dfrac{45}{t}=1 \\ \dfrac{240}{s}-\dfrac{240}{t}=3 \end{cases}$$

leads to

$$\frac{1440}{s}=25 \text{ or } s=\frac{1440}{25}=\frac{288}{5}=57\frac{3}{5}\text{ weeks } t=\frac{1440}{7}=205\frac{5}{7}\text{ weeks, and}$$

$$f=\frac{1440}{23}=\frac{288}{5}=62\frac{14}{23}\text{ weeks}$$

needed for each company to complete the job if it works alone.

Let x be the number of weeks the three firms need to complete the job if they work together, then

$$\frac{x}{f}+\frac{x}{s}+\frac{x}{t}=1 \Rightarrow x\cdot\left(\frac{1}{\frac{1440}{23}}+\frac{1}{\frac{288}{5}}+\frac{1}{\frac{1440}{7}}\right)=1 \Rightarrow x=\frac{288}{11}=26\frac{2}{11}\text{ weeks.}$$

Three firms can finish maintenance in $26\frac{2}{11}$weeks.

Remarkable Tunnels. Let 1 denote the entire job and f, s, and t be the number of weeks required by the first, second, and third firm to complete the job if a firm works alone. Then $1/f$, $1/s$, and $1/t$ are the productivities of each firm, and, considering the number of weeks they need if they work in pairs, we get the system

$$\begin{cases} \dfrac{30}{f}+\dfrac{30}{s}=1 \\ \dfrac{48}{f}+\dfrac{48}{t}=1. \\ \dfrac{45}{s}+\dfrac{45}{t}=1 \end{cases}$$

Adding three equations of the system and collecting like terms we get

$$\frac{78}{f}+\frac{75}{s}+\frac{93}{t}=3, \text{or} \frac{26}{f}+\frac{25}{s}+\frac{31}{t}=1.$$

The first, second, and third firms will be hired for the 26, 25, and 31 weeks correspondingly.

The Blue Lagoon. Let x and y be the amount of water or the productivity of each source per hour and 1 be the entire job to fill in the spa. If two sources are open then both can fill the spa in 12 hours or $12(x + y) = 1$. Next, $1/x$ and $1/y$ are the time or number of hours needed for each source to fill the spa if only one of them is open. The time during which the second source fills two third of the spa alone is $2/3 \cdot 1/y$ hours. Then $2/3 \cdot 1/y \cdot x$ is the portion of the spa filled by the first source during this time. Similarly, $3/11 \cdot 1/x \cdot y$ is the portion of the spa filled by the first source during the time the first source can fill 3/11 of the spa. Hence,

$$\begin{cases} 12(x+y)=1 \\ \frac{2}{3}\frac{1}{y}x+\frac{3}{11}\frac{1}{x}y=1-\frac{3}{22} \end{cases} \Rightarrow \begin{cases} y=\frac{1}{12}-x \\ \frac{2}{3}\cdot\frac{x}{\frac{1}{12}-x}+\frac{3}{11}\cdot\frac{\frac{1}{12}-x}{x}=\frac{19}{22} \end{cases} \Rightarrow \begin{cases} y=\frac{1}{12}-x \\ 952x^2-62x+1=0 \end{cases}.$$

Solving the quadratic equation

$$952x^2-62x+1=0 \text{ as } x=\frac{31\pm\sqrt{31^2-952}}{952},$$

we get $x=\frac{1}{28}$ and $x=\frac{1}{34}$ that leads to the solution

$$\begin{cases} x=\frac{1}{28}, \; y=\frac{1}{21} \\ x=\frac{1}{34}, \; y=\frac{11}{204} \end{cases}.$$

The problem has two solutions: 28 and 21 hours or 34 and 204/13 hours, but only the first pair is in integer numbers.

Construction Projects. Let 1 denote the completion of the entire maintenance and x be the number of days to provide the maintenance by the first company alone. Then $2x$ and $(x + 2)$ are the number of days to complete the maintenance by the second and third companies, and

$$\left(\frac{1}{x}+\frac{1}{2x}+\frac{1}{x+2}\right)=\frac{13}{7}\left(\frac{1}{2x}+\frac{1}{x+2}\right)\Rightarrow 35x+42=39x+26\Rightarrow x=4.$$

Thus, the first company needs 4 days, the second one requests 6 days, and the third one asks for 8 days to complete the maintenance.

Let us use m for requested funding by the first company and compare the amounts requested by each company.

Comparison of estimates from the first and third companies, that is $4m$ and $8(m - 20{,}000)$, leads to

$8(m - 20{,}000) - 4m = 4m - 160{,}000$, which is

- less than zero if $m < 40{,}000$, that means hiring the third company is cheaper;
- equals 0 if $m = 40000$, that means the first and third companies' financial estimates are the same;
- greater than 0 if $m > 40{,}000$, that means hiring the first company is cheaper.

Comparing the first and second companies' estimates we get:

- $4m$ is always greater than $6(m/2)$. Thus, out of these two companies it is profitable to hire the second one.

Finally, comparing the second and third companies' financial requests, that is $8(m - 20{,}000)$ and $6(m/2)$, we obtain: $8(m - 20{,}000) - 3m = 5m - 160{,}000$, which is

- less than zero if $m < 32{,}000$, that means hiring the third company is cheaper;
- equals 0 if $m = 32{,}000$, that means the second and third companies' estimates are the same;
- greater than 0 if $m > 32{,}000$, that means hiring the second company is cheaper.

It is profitable to make the following decision:

- hire the third company if $m < 32{,}000$,
- hire the second or the third company if $m = 32{,}000$,
- hire the second company if $32000 < m < 40{,}000$,
- hire the first or the second company if $m = 40{,}000$,
- hire the second company if $m > 40{,}000$.

Sources

www.atlasobscura.com/places/jemaa-elfna

www.bluelagoon.com/

en.wikipedia.org/wiki/Blue_Lagoon

en.wikipedia.org/wiki/Jemaa_el-Fnaa

ich.unesco.org/en/RL/cultural-space-of-jemaa-el-fna-square-00014

www.smithsonianmag.com/smithsonian-institution/unexpected-history-air-conditioner

www.footstepsontheglobe.com/destinations/europe/spain/barcelona/15-awesome-facts-about-sagrada-familia-that-youll-love/

en.wikipedia.org/wiki/Sagrada_Fam%c3%adlia

www.delish.com/food/g23928887/spas-with-beer-baths/

www.reuters.com/article/us-france-castle-idUSKCN11K1XJ

en.wikipedia.org/wiki/Haiti

demandafrica.com/travel/destinations/fifteen-amazing-facts-you-never-knew-about-haiti/

www.ontheroadtrends.com/longest-tunnels-on-the-world

constructionreviewonline.com/biggest-projects/10-of-the-largest-construction-projects-in-the-world/

graconllc.com/largest-construction-projects-in-history/

3.3 Speed and Distances

Sailing in the Caribbean

The closely located to each other Caribbean islands stretch like a long 2500-mile chain. Over 44 million people of indigenous, African, and European descent live in Antigua and Barbuda, Bahamas, Belize, Costa Rica, Cuba, Dominica, Guatemala, Saint Kitts and Nevis, and 13 other independent Caribbean countries and 15 dependencies. They mostly speak English, French, Dutch, and Spanish. The Caribbean is rich in culture and history and has a fantastic climate that attracts tourists. Sailors can cruise the Caribbean for months and anchor up to discover its magnificent landscapes and unique biodiversity.

The sail distance between Caribbean islands is provided in nautical miles used in air, marine, and space navigation, 1 nm = 1.852 km, and the related unit of speed is the knot, one nautical mile per hour.

Problem. The distance between Basseterre in Saint Kitts and Nevis) and Falmouth in Antigua and Barbuda is 62 nm and from Falmouth to Deshaies in Guadeloupe is 43 nm. A cruiser sails from Basseterre to Falmouth at 5 knots and then from Falmouth to Deshaies. How fast should the cruiser sail from Falmouth to Deshaies to have an average cruising speed of 9 knots for the entire Basseterre–Falmouth–Deshaies sailing?

Answer. Impossible

Explore Dominica

The Commonwealth of Dominica is a mountainous Caribbean Island country with natural hot springs, amazing beachies, unique waterfalls, volcanic lakes, and tropical rainforests. The area of the country is only 290 sq mi. English is its official language. Roseau is the Dominica's capital and largest city. French planters settled in Dominica before the British captured the island in 1759. The island was recaptured by French and then again by British. Dominica achieved full independence on November 3, 1978.

Problem. To explore Dominica, a cyclist left Saint Joseph for Portsmouth and a hiker left Portsmouth for Saint Joseph. What would be the distance between them 30 minutes before they meet? Use the minimum number of hints below to answer the question.

A. The distance between Saint Joseph and Portsmouth is 21 km.
B. The cyclist left Saint Joseph one hour after the hiker left Portsmouth.
C. The cyclist bikes 12 km per hour and the hiker walks 4 km per hour.

Answer. 8 km, C

The Middle East

Algeria, Bahrain, Egypt, Iran, Iraq, Israel, Jordan, Kuwait, Morocco, Oman, Qatar, Tunisia, Yemen, and United Arab Emirates are among 18 countries of Arabian Peninsula, West Asia, and Nort-East Africa that make up the Middle East. Most of them, including two neighboring Middle Eastern countries, Lebanon, and Palestine, are part of the Arab world. Their rich history goes back to ancient times. Their bright heritage has been recognized for millennia. This place is considered as the cradle of our civilization and the birthplace of Judaism, Christianity, and Islam as well as the birthplace of several other worldwide religions. The borders between the countries have varied dramatically. The State of Palestine is now between Israel, the West Bank, and the Gaza Strip. The city of Ramallah is its de facto capital, though most Palestinians consider Jerusalem as their capital. Three Christmas holidays are celebrated in Palestine. The Republic of Lebanon is at the crossroads of the Mediterranean Basin and the Arabian Peninsula. Its capital Beirut is just 20 miles from the world's oldest continuously inhabited city Byblos.

Interestingly, the world *papyrus* in Greek is *Byblos* from which the English word *Bible* was originated.

Problem. Use just one of the following hints to find the driving distance between Beirut and East Jerusalem. Explain why other hints are not necessary to find a solution.

i. The straight line or flying distance from East Jerusalem to Beirut is 168 miles.
ii. The driving distance is reduced from ¼ to ½ within 1 hour.
iii. The average speed to drive from Beirut to East Jerusalem is between 50 and 60 mph.
iv. Driving 51 miles decreases the driving distance from ¼ to ½.

Answer. 204 mi

A Salesperson's Problem

Western Africa of Sub-Saharan Africa is the westernmost region of Africa. Mali, Guinea, Burkina Faso, and Cote d'Ivoire are among its 16 countries. Being rich with gold, diamonds, and other minerals and natural resources, Western Africa is the fastest growing region demographically and economically on the African continent.

Problem. A travelling salesperson wants to visit four Western African cities, Sikasso in Mali, Bobo-Dioulasso in Burkina Faso, Kankan in Guinea, and Korhogo in Cote d'Ivoire where he run a business. None of three cities are on a straight line. Neither the order of cities to visit nor a route is important to him. To choose the shortest way, the merchant finds the distances between the cities and prepares a graph shown in Figure 3.1. What route do you recommend for the salesperson to choose?

Answer. None

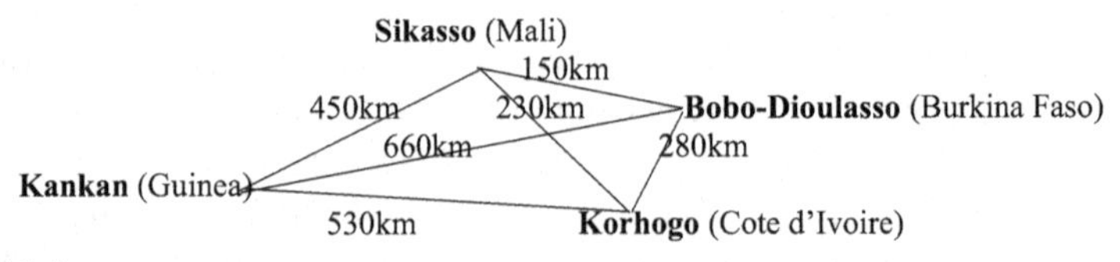

FIGURE 3.1

Visit Malta

Malta is surrounded by crystal clear waters and has the world's best climate and the best place to retire. Catholicism is its official religion. Being ruled by the Phoenicians, Carthaginians, Romans, Byzantines, Arabs, Normans, Sicilians, Spanish, Knights Templars, French, and British before gaining its independence in 1964, Malta thoughtfully preserves its heritage and relics including the man-made structures that are older than the pyramids of Egypt, the Great Wall of China, and Stonehenge in England. Being among the world's smallest countries and most densely populated countries, Malta is often referred to as a *city-state*.

Valletta has been the capital since 1571, taking this title from Birgu renamed as Cittá Vittoriosa and the oldest city of Mdina. This smallest EU capital is one of the most concentrated historical areas and is listed as a UNESCO World Heritage Site. Vittoriosa, Senglea and Cospicua are called *the three cities* of Malta due to their strategic importance.

Problem. Three friends, a cyclist from Vittoriosa, a hiker from Cospicua, and a jogger from Valletta have arranged their meeting at the ancient city of Mdina. The distances between Valletta and Mdina and between Vittoriosa and Mdina are the same and equal to 14 km, while the distance between Cospicua and Mdina is 1 km less. The cyclist is planning to leave Vittoriosa at noon. He is biking 14 km/h. The jogger is running 8 km/h while the hiker walks 5 km/h. What time should a jogger and a walker leave their cities for the friends to come to Mdina at the same time?

Answer. 11:15, 10:24

Sunset in Panama

The Panama Canal that connects Atlantic and Pacific Oceans is one of the modern *seven wonders* of the world. It was built by the US Army Corps of Engineers between 1904 and 1914 after the French failed to construct it following the design of the Suez Canal in Egypt. Panama is famous not only for having this famous canal but also for the possibility of watching the sunrise and sunset over two different oceans from the same spot at the county's narrowest width.

Panama City is the only capital in the world with a rainforest. Interestingly, Panama celebrates two independence days, from Spain in 1821 and from Colombia in 1903. Finally, this country is one of the least populated in Central America and in the world.

Problem. To check whether both sunset and sunrise can be seen not only from the middle of the narrowest part of Panama but also from its both endpoints, two cyclists, Alex and Neel, decided to watch the sunset from two Atlantic and Pacific Ocean coastline points. It takes 5 hours for Alex to bike the distance from Atlantic to Pacific coastlines and 3 hours and 20 minutes for Neel to cover the same distance. Because Neel chose to bike toward the Atlantic coastline, the cyclists agreed to start from the same point that is 10 miles closer to the Atlantic coastline than to the Pacific coastline to arrive at the end point at the same time. What is the length of the narrowest part of Panama?

Answer. 50 mi

Camino de Santiago

It is believed that remains of Saint James the Great were carried by boat from Jerusalem, where he was beheaded by order of King Agrippa, to Galicia in northwesternmost Spain and were laid to rest on the spot where the magnificent Cathedral of Santiago de Compostela was built, and the city of Santiago de Compostela was founded later. The pilgrimage to *Santiago de Compostela* or the *Way of St. James* has become one of the most important Christian pilgrimages alongside with the Via Francigena to Rome in Italy and the pilgrimage to Jerusalem. Depending on a starting point, the Camino de Santiago or *Pilgrimage of Compostela* has several pilgrimage routes, but all of them lead to the shrine of Santiago. Camino Frances,

Camino Portugues, Camino Primitivo, Camino Finisterre, and other pilgrimages are well developed now with pointers that show the direction and remaining distance, places to stay from cheap albergues to comfortable hotels, numerous cafes, and places to relax. The backpack of pilgrims is decorated with a scallop shell as proof of their journey. Pilgrims are like a family; they help and support each other independently of their race, age, heritage, and religion. The Santiago de Compostela was declared as the first European Cultural Route in 1987 and UNESCO's World Heritage Site in 1993.

The Original Way through Galicia is the very first of the pilgrimage route to Santiago. It crosses a spectacular mountain range. On the other side, the Portuguese Way is not that challenging but also scenic and fascinating.

Problem. Two brothers decided to take two different pilgrimage routes and then share their experience. Andrew chose the Original Way, while Phillip decided to walk the Portuguese Way, which is 138 km longer. They calculated that, in order to arrive in Santiago de Compostella at the same day, Phillip should cover 6 km more each day. After walking for 17 days, Anderew took one day of relaxation to tour a beautiful city. Because of this, he had to walk 4 km more each remaining day to arrive in Santiago de Compostella as planned. How long are the Original and the Portuguese Ways?

Answer. 460 km, 598 km

The Flattest Countries

Maldives is an archipelagic state in the Indian Ocean. It is the lowest and flattest country in South Asia and in the world, for its maximum elevation above sea level is just 2.4 meters. With the highest elevation of 300 meters, Lithuania is the European flattest country, while Gambia with the highest elevation of 34 meters is the African flattest country. Australia is the flattest continent in the world with the most of its land below 600 meters above sea level. It is also called the *Desert Continent* because much of its land is in the desert. Therefore, most of its population live along the coast in Sydney, Melbourne, Brisbane, Perth, and Adelaide connected by scenic well-developed roads.

Problem. A car leaves Sydney for Melbourne at a speed of 55 mph. Fifteen minutes later another car leaves Sydney for Melbourne at 66 miles per hour. How far from Sydney will the cars be when the second car catches up with the first? What is the driving distance between Sydney and Melbourne if the second car should continue driving for 6 hours and 55 minutes more before reaching Melbourne?

Answer. 2 h, 572 mi

Cities with the Most Airports

Air travel between cities is becoming more and more important. Hounslow Heath Aerodrome in England was the world's first airport to operate international commercial aviation. It was opened in 1919. Thousands of airports have been built since that time. Many cities have more than one airport. Los Angeles has five airports close to the city. Four international and domestic airports are in Tokyo, Moscow, and Stockholm. With six commercial airports, London and New York lead the list of cities with the greatest number of airports with London having the highest passenger volume. London Heathrow (the largest) and Gatwick are its largest airports.

Problem. The taxi fare from London Heathrow Airport to Gatwick Airport is 4 British pounds for the first half kilometer and 0.38 British pounds for the next 0.4 km that added up to 61 British pounds for a ride. What is the distance between Heathrow and Gatwick Airports?

Answer. 60.5 km

Balkan Countries

North Macedonia and Bulgaria are not only sharing the border but also have a lot in common. Both Balkan countries value their rich heritage, enjoy delicious but simple and healthy food, love dance, and claim to be the birthplace of Cyrillic script. They are NATO members. In addition to the euro, both countries have their official currency, the Bulgarian lev and Macedonian denar. Being Orthodox Christian countries, they have a great number of churches, monasteries, and other religious constructions. Macedonian city Ohrid with 365 churches, Boyana Church and the tenth century Rila Monastery in Bulgaria are among UNESCO World Heritage Sites. Built to commemorate the second millennium anniversary of Christianity in Macedonia, the Millennium Cross of Vodno Mountain is visible from any point of its capital Skopje. The Bulgaria capital, Sofia, originally called Serdika, was renamed after St. Sophia Cathedral in the 14th century.

Natural sites of the two countries are also worth visiting. Bulgarian shores of the Black Sea, its Rose Valley in the Rhodope Mountains, and natural mineral water springs are so welcoming. The Ohrid Lake in North Macedonia is among the deepest European lake and the world's oldest ones.

Throughout history, North Macedonia was a part of the Alexander, Roman, Byzantine, and Ottoman Empires, before gaining its independence in 1991 as one of the successor states of Yugoslavia. The country was accused by Greece for using the Vergina Sun, Alexander the Great, and other symbols and figures historically considered to belong to Greek culture. After 10 years of a severe dispute between Greece and the Republic of Macedonia, the name of the later was changed to the Republic of North Macedonia despite protests of its citizens. With over 30 mountaintops above 2000 meters, this Balkan country is the world's second most mountainous country after Montenegro.

Problem. One car left Skopje for Sofia and the second car left Sofia for Skopje one and a half hours later. Both cars drove for an hour before the distance between them became 42 km. The second car left Sofia 1 hour and 30 minutes before the first car left Skopje, and then both cars drove for an hour before the distance between them became 60 km. The first car arrived in Sofia after 4 hours of driving. How fast were both cars driving? What is the distance between Skopje and Sofia?

Answer. 60 km/h, 48 km/h, 240 km

Solutions

Sailing in the Caribbean. It takes a cruiser 12.4 h = 62 nm / 5 knots to sail from Basseterre to Falmouth that is already greater than expected 11.67 h = 105 miles / 9 knots to sail the entire distance of 105 nm from Falmouth to Deshaies with 9 knots.

Explore Dominica. The cyclist and hiker cover 16 km per hour. It means they will have 8 km between them 30 minutes before they meet.

The Middle East. The driving distance $\frac{1}{2}-\frac{1}{4}=\frac{1}{4}$ or one quarter is related to 51 miles by Hint iv. Therefore, the total distance is $51 \cdot 4 = 204$.

The driving distance between Beirut and East Jerusalem is 204 miles.

A Salesperson's Problem. Recommendation: Check all calculations and correct the graph-map. There is a mistake. Consider a triangle with end points at Sikasso, Bobo- Dioulasso, and Kankan. The sum of its two sides is less than the third side, $450 + 150 < 660$. Such a triangle does not exist. These three cities cannot be on one line either because $450 + 150 = 600 \neq 660$.

Visit Malta. Since the distance between Vittoriosa and Mdina is 14 km and the cyclist is biking 14 km/h, he will reach Mdina in one hour, that is, at 1:00 pm. The hiker needs $13/5 = 2.6$ h or 2 h and 36 min to walk from Cospicua to Mdina at 5 km/h. It means, he has to leave Cospicua 1 h and 36 min before noon, that is, at 10:24. The jogger needs $14/8 = 1.75$ h or 1 h and 45 min to cover the distance between Valletta and Mdina walking 5 km/h. Thus, he has to leave Valletta 45 min before noon at 11:15.

A jogger should leave Cospicua at 11:15 am and a walker should leave Valletta at 10:24 am.

Sunset in Panama. Let u be the speed of Alex, and v be the speed of Neel. It takes them 5 hours and 3 hours and 20 minutes or $3\frac{1}{3}=\frac{10}{3}$ hours, respectively to bike from one coastline to another. The distance between coastlines is $\frac{10}{3}u = 5v$, from which $v=\frac{2}{3}u$ follows. The half of the distance is $\frac{10}{3\cdot 2}u=\frac{5}{3}u$. The cyclists started 10 miles closer to the Atlantic coastline, that is, 5 miles away from the middle point. Hence, Alex biked $\frac{5}{3}u+5$ and Neel biked $\frac{5}{3}u-5$. They came to coastlines at the same time, that is,

$$\frac{\frac{5}{3}u+5}{u}=\frac{\frac{5}{3}u-5}{v}, v=\frac{2}{3}u.$$

Therefore, $u = 15$ and the total distance is $15 \cdot 10/3 = 50$.

The narrowest length of Panama that separates two oceans is just 50 miles or 80 kilometers.

Camino de Santiago. Let s km, $s \neq 0$, be the length of the Original Way and v km/day, $v \neq 0$, the number of kilometers Andrew was supposed to cover each day (his speed per day), then $(s + 138)$ km is the length of the Portuguese Way, $(v + 6)$ km/day is the number of kilometers Phillip was supposed to cover each day (his speed per day). The brothers planned to arrive in Santiago de Compostella at the same time, it means they should walk the same number of days or

$$\frac{s}{v} = \frac{s+138}{v+6}.$$

Since Anderew took one day of relaxation after 17 days of walking covering the distance of ($s - 17v$) km, then, to catch up and arrive in Santiago de Compostella as planned, he started to walk 4 km more each remaining day.

$$\frac{s}{v} = 17+1+\frac{s-17v}{v+4}.$$

The two equations produce the system of nonlinear equations

$$\begin{cases} \frac{s}{v} = \frac{s+138}{v+6} \\ \frac{s}{v} = 18+\frac{s-17v}{v+4} \end{cases}.$$

The first equation leads to $sv + 138v = sv + 6s$, which gives $s = 23v$. The second equation is transformed to $sv + 4v = 18v^2 + 72v + sv - 17v^2$, which with the substitution of $s = 23v$ leads to $v(v - 20) = 0$, that has two solutions, $v = 0$ and $v = 20$. The first one does not satisfy the problem. Therefore, $s = 20 \cdot 23 = 460$.

The Camino Primitivo is 460 km, while the Camino Portugués is 598 km.

The Flattest Countries. Let t hours be the driving time of the first car before the second car catches up. The second car drives 15 minutes or $(t - 1/3)$ hours less. Both cars drive the same distance. Then

$55t = 66(t - 1/3) \Rightarrow t = 2.$

Two cars were $2 \cdot 55 = 110$ miles from Sydney.

The second car should drive 6 hours and 55 minutes more before reaching Melbourne, that is, the driving time is 8 hours and 55 minutes without 15 minutes that with the speed of 66 miles per hour lead to the driving distance of 572 miles.

The driving distance between Sydney and Melbourne is 572 miles.

Cities with the Most Airports. A 0.5-km ride costs 4 British pounds. The remaining 57 British pounds cover

$$57 \cdot \frac{0.4}{.38} = 60\text{km}.$$

Thus, the distance between Heathrow and Gatwick Airports is 60 + 0.5 = 60.5 km.

Balkan Countries. Let u km/h and v km/h be the average speed of the cars leaving Skopje and Sofia, then $4u$ is the distance between Skopje and Sofia. The statement that the first car left Skopje one and a half hours before the second car left Sofia, and then both cars drove for an hour before the distance between them became 42 km can be written as

$$(1+1\frac{1}{2})u+1v=4u-42.$$

The statement that the second car left Sofia one and a half hours before the first car left Skopje, and then both cars drove for an hour before the distance between them became 60 km can be modelled as

$$1u+(1+\frac{1}{2})v=4u-60.$$

Both equations combined lead to the system

$$\begin{cases}(1+1\frac{1}{2})u+1v=4u-42\\ 1u+(1+1\frac{1}{2})v=4u-60\end{cases}\Rightarrow\begin{cases}\frac{3}{2}u-v=42\\ 3u-\frac{5}{2}v=60\end{cases}\Rightarrow u=60, v=48, 4u=240.$$

The distance between Skopje and Sofia is 240 km.

Sources

improvesailing.com/destinations/sailing-time-between-all-caribbean-islands-table-tips

www.worldatlas.com/articles/caribbean-countries.html

en.wikipedia.org/wiki/Dominica

www.google.com/search?q=distance

www.britannica.com/place/Palestine

en.wikipedia.org/wiki/State_of_Palestine,

en.wikipedia.org/wiki/Lebanon

www.enjoytravel.com/us/travel-news/interesting-facts/interesting-facts-lebanon,

en.wikipedia.org/wiki/Middle_East

www.google.com/search?q=distance,en.wikipedia.org/wiki/West_Africa

brbgonesomewhereepic.com/20-facts-about-malta/

www.google.com/search?q=distance

theculturetrip.com/central-america/panama/articles/11-fascinating-things-you-didnt-know-about-panama/

followthecamino.com/en/

www.pilgrim.es/en/routes/

www.guinnessworldrecords.com/world-records/flattest-country

byjus.com/question-answer/why-is-australia-called-the-flattest-continent

simpleflying.com/the-cities-that-have-the-most-commercial-airports/

www.aqtsolutions.com/what-was-the-worlds-first-airport/

www.google.com/search?q=distance

www.chasingthedonkey.com/fun-facts-about-macedonia-facts/

nomadsunveiled.com/facts-on-bulgaria/

3.4 From Capital to Capital

Island Countries

Madagascar is 400 kilometers off the coast of Southeast African country Mozambique across the Mozambique Channel. It left Africa over150 million years ago and split from India later. Being isolated from continents for so many million years, Madagascar has acquired the unique flora and fauna that cannot be found anywhere else. Lemurs, zebu, periwinkle to cure cancer, and many other animals and plants can be found in the wild. Madagascar is the fourth largest island, and the Republic of Madagascar is the second largest island country in the world. Its capital is Antananarivo. The first Malagasies arrived from Asia, Africa, and later from Europe sharing their heritage and forming a diverse country. The island used to be heaven for European pirates.

Mozambique is famous for its beaches, crystal water of Indian Ocean, coral islands of the Quirimbas Archipelago, reefs of the Bazaruto Archipelago, marine parks, ancient ruins, and traditional music. It borders Tanzania, Zambia, Zimbabwe, Malawi, Eswatini, and South Africa. Its capital is Maputo. The country has just two seasons and two geographic regions.

Problem. It takes 3 hours for a plane to cover the distance from Maputo to Antananarivo. Because of a 20-minute delay, the plane flew 75 km/h faster to arrive in Antananarivo on time. What is the flying distance between Maputo and Antananarivo?

Answer. 1800 km

Central European Capitals

Hungary and Romania in Central Europe are neighbors. *Romania* got its name from the Latin word *Romanus* meaning *citizen of the Roman Empire*. A legend states that its capital *Bucharest* was named after a shepherd Bucur who founded the city. Its nickname *Little Paris* came after building Arcul de Triumf there like the Arc de Triomphe in Paris. The 84-m high Parliament Palace is the heaviest building in the world and the second largest after the Pentagon in the US. Romania is the birthplace of four Nobel Prize laureates and the inventor of the first espresso machine Francesco Illy. The country was the first in Europe to have electric street lighting in 1889.

Hungary claims to have the world's first official wine region in Tokaj. Three cities, Buda, Obuda, and Pest, merged in 1873 to form Budapest that became the capital of this landlocked country. The country treasures its hundreds of thermal mineral spas. Its old underground is listed as a UNESCO World Heritage Site. The Hungarian language is unique and is hardest languages to learn. The country is proud of achievements of its citizens like the discovery of vitamin C by Albert Szent-Gyorgyi in 1930 for which he earned the Nobel Prize and the Rubik's Cube invented by Erno Rubik in 1974. Paprika is a symbol of Hungarian cuisine and palinka is its most popular beverage, which is believed to cure all diseases.

Problem. A couple spent four days driving from Budapest to Bucharest and made four stops to see their relatives and friends. During the first day they covered 1/40 of the driving distance between Budapest and Bucharest. They doubled that distance and drove 2 kilometers more on the second day. The couple tripled the distance of the previous day and covered 3 kilometers more on the third day. On the fourth day they quadrupled the distance of the third day and drove 100 kilometers more to reach Bucharest. What is the driving distance between Budapest and Bucharest?

Answer. 840 km

The Furthest Closest Capitals

What is the capital city of Australia? Sidney? Oh, no! It is Canberra with the name derived from *Kanbarra* meaning a *meeting place* for the local Ngabri people.

Now, try to name the capital of New Zealand. You cannot? It is Wellington! The city was named after Arthur Wellesley, the first Duke of Wellington and the hero of the Battle of Waterloo. These two capitals of sovereign countries have the longest distance between two capitals closest to each other. Moreover, Wellington and Canberra are much closer to each other than to the capital of any other country. It means they are furthest away from the closest capital. The capital of New Zealand also shares another record with the capital of Spain, Madrid. Located almost opposite each other on the globe, these two capitals are farthest away from each other.

Problem. Assume that there is a magic ground transportation highway that connects two capitals, Wellington and Canberra, following a straight line. How far are these capitals from each other if the 500 miles of that distance could be covered during the first period, 1/7 of the entire distance could be covered during the second period, and a half of the entire distance could be covered during the last third period of time?

Answer. 1400 miles

Multiple Capitals

The capital of a country is its main city that holds the primary status within a country and where the royals or government are usually located. It is not always the largest or the most famous city of a country. Some countries have several capitals. With Cape Town, Pretoria, and Bloemfontein, South Africa has the most national capital cities in the world. Although Sucre is the constitutional capital of Bolivia, La Paz is its de facto capital, the home to its government offices and its financial center. It is also the world's highest capital city. El Alto International Airport in La Paz is the world's highest international airport and the highest commercial airport outside of China.

Problem. Assume that there is an imaginary straight line that connects two capital cities. How far are these capitals if a mi of that distance is covered during the first period of time, $1/n$ of the entire distance is covered during the second period of time, and $1/m$ of the entire distance is covered during the third period of time?

Does this problem have a solution for any a, m, and n, $anm \neq 0$?

Answer. $\dfrac{anm}{nm-n-m}$, no

Capitals of Bordering Countries

Several pairs of capital cities of bordering countries can compete to be furthest apart from each other. Among them are Paris and Brasília, capitals of France and Brasilia. Surprised? Indeed, one of them is a European capital and another one is a South American capital. How is it possible that they have a joint border if they are located on different continents? Yes, they do, because French Guiana that shares the border with Brazil is an overseas department of France.

Russia and North Korea share a border. Their capitals, Moscow and Pyongyang, are furthest away compared to capitals of all countries that share the land border. The distance between them is even greater than the distance between the two most distant provincial capitals in the world's second largest country by area, Canada, or the length of the world's longest country, Chile.

Problem. The straight distance between Moscow and Pyongyang can be split into four nonzero sections. The first section is 4823 kilometers, the second one is the eighths of the entire distance, the third portion is the ninths of the entire distance, and the last one is eighty ninths of the entire distance. How far is Moscow from Pyongyang?

Answer. 6408 km

Popular Capitals

The capital of a country is a place where its government meets delegations from other countries, and it is also the #1 tourist destination. The capital of Thailand, Bangkok, is the most visited capital in the world followed by the capital of England, London. Citizens of the capital of Italy, Rome, are very stylish, so tourists should dress nicely trying not to look weird. Remember, *when in Rome, do as the Romans do.*

Problem. The straight distance between two capitals can be split if four non-zero sections. The first section is a miles, the second one is a $1/n$ part of the entire distance, the third portions is a $1/m$ part of the entire distance, and the last one is a $1/l$ part of the entire distance. How far are the capitals from each other? Does the problem have a solution for any a, l, m, n and n, $alnm \neq 0$?

Answer. $\dfrac{anml}{nml - ml - nl - mn}$, no

The Closest Capitals

National capital cities are spread around the globe. The US capital, Washington, is 460 miles away from its nearest Canadian capital, Ottawa. The closest capitals of two sovereign countries are Vatican City of Vatican and Rome of Italy. Indeed, Vatican City is located inside Rome. Two second closest capitals and the closest capitals between sovereign countries (indeed, the Vatican is a city-state!) are Kinshasa of Democratic Republic of the Congo and Brazzaville of Republic of the Congo located on different banks of the mighty Congo River. Since these two African countries do not have a bridge over the river, people should take a five-minute flight or a boat ride.

Problem. A road-and-rail bridge between Kinshasa and Brazzaville has been discussed for years. After its construction, the length of the bridge would be 1 mile in addition to 1/5 of its total length, one quarter of the rest of the length, and ¾ of the mile. How long would be the bridge over the Congo River?

Answer. 2.5 mi

Exceptionality of Capitals

Capital cities are located in different climate zones. Reykjavik, the capital of Iceland, is the world's most northerly capital and the European most westerly capital city, while Wellington of New Zealand is the southernmost capital in the world.

Cape Town of South Africa is the most biodiverse capital city and is home to many endemic species. The Japanese capital, Tokyo, is the most populous city and the most natural disaster-prone city in the world. Cairo, the capital of Egypt, is the driest capital city, while Monrovia in Liberia is the wettest capital city.

Problem. The straight distance between two capitals can be split in 4 nonzero sections. The first section is a miles, the second one is a $1/n$ part of the entire distance, the third portions is a $1/m$ part of the distance that is left, and the last one is b miles, $abnm \neq 0$. How far apart are the capitals from each other?

Answer. $\dfrac{(bm+a(m-1))n}{nm-m-n+1}$

Capitals Named after Presidents, Kings, and Tribes

Many buildings, squares, cities, and capitals of a country are named after kings, presidents, and other famous people and tribes. The French capital, Paris, was named after the Parisii tribe and Ecuador's capital is called Quito after the Quitu tribe. The capital city of Brunei was named Bandar Seri Begawan after the Sultan's late father. The last king of the Bubi, Malabo Lopelo Melaka, is remembered in the Equatorial Guinea capital's name, Malabo.

Madison in Wisconsin, Lincoln in Nebraska, Jefferson City in Missouri, and Jackson in Mississippi are four state capitals named after US presidents. The capitals of two countries have been named after American presidents. The USA capital, Washington DC, took the name of the first president, and the capital of Liberia, Monrovia, was named after the fifth president of the United States.

Problem. Two capitals, American Washington DC and Liberian Monrovia, are connected with a straight road that lies entirely on land. The road is divided into four segments of nonzero length. The first segment is 779 miles, the second one is half of the entire length, the third segment is the third of the entire length, and the fourth segment is a quarter of the remaining length. What is the straight land distance between Washington DC and Monrovia? How long is each segment?

Answer. No solution

Multilingual Capitals

Luxembourg, the capital of Luxembourg, is the most multilingual city. Equatorial Guinea, capital city Malabo, is the only African country with Spanish as its official language, while Guyana with the capital city Georgetown is the only South American country with English as its official language, though most of its population speak Guyanese Creole. Cairo, the capital of Egypt, is the only city in the world that has one of eight ancient wonders of the world, the Great Pyramid at Giza, still left standing.

Problem. Let the distance between two capitals be divided into three segments of nonzero length. The first segment is of a miles, the second one is a $1/n$ part of the entire distance, and the third segment is a $1/m$ part of the remaining distance, $m > 1, n > 1$. What is the distance between capitals?

Answer. No solution

Solutions

Island Countries. Let v km/h be the average speed of the plane and (v + 75) km/h be its adjusted speed due to a 20-min or 0.25-hour delay. The same distance is covered, that is,

$$3v = \frac{8}{3}(v+75) \text{ or } 9v = 8v + 600, v = 600.$$

The flying distance between the two capitals, Maputo and Antananarivo, is 1800 km.

Central European Capitals. 1st way. Let x km be the distance covered during the first day. Then $40x$ km is the distance between Budapest to Bucharest and

$$x+(2x+2)+(3(2x+2)+3)+(4(3(2x+2)+3)+100)=40x,$$

where each term is related to one day of the travel. Then $33x+147=40x$or $x=21$. Hence, $40x$ is $40{\cdot}21=840$.

2nd way. Let x km be the total distance, then

$$x=\frac{1}{40}x+\left(2\cdot\frac{1}{40}x+2\right)+\left(3\cdot\left(2\cdot\frac{1}{40}x+2\right)+3\right)+4\cdot\left(3\cdot\left(2\cdot\frac{1}{40}x+2\right)+3\right)+100$$
$$\Rightarrow x=840.$$

The driving distance between Budapest and Bucharest is 840 kilometers.

The Furthest Closest Capitals. Let x mi, $x\neq 0$, be the distance between Wellington and Canberra, then

$$x=500+\frac{1}{7}x+\frac{1}{2}x\Rightarrow x-\frac{1}{7}x-\frac{1}{2}x=500\Rightarrow\frac{14-2-7}{14}x=500$$
$$\Rightarrow x=\frac{500\cdot 14}{5}=1400.$$

The largest distance between Wellington and Canberra, is 1400 mi.

Multiple Capitals. Let x mi, $x\neq 0$, be the distance between the capitals, then

$$a+\frac{1}{n}x+\frac{1}{m}x=x\Rightarrow x-\frac{1}{n}x-\frac{1}{m}x=a\Rightarrow x\frac{nm-m-n}{nm}=a\Rightarrow x=\frac{anm}{nm-n-m}.$$

The problem does not have a solution if $nm\leq n+m$, because then the distance becomes negative.

Capitals of Bordering Countries. Let x km be the distance between Moscow and Pyongyang, then

$$x=4823+\frac{1}{8}x+\frac{1}{9}x+\frac{1}{89}x\Rightarrow x-\frac{1}{8}x-\frac{1}{9}x-\frac{1}{89}x=4823$$
$$\Rightarrow\frac{8\cdot 9\cdot 89-9\cdot 89-8\cdot 89-8\cdot 9}{8\cdot 9\cdot 89}x=4823$$

$$\Rightarrow x=6408.$$

The Russian and North Korean capitals, Moscow and Pyongyang, are 6408 km apart.

Popular Capitals. Let x be the distance between two capitals, then

$$a+\frac{1}{n}x+\frac{1}{m}x+\frac{1}{l}x=x\Rightarrow x-\frac{1}{n}x-\frac{1}{m}x-\frac{1}{l}x=a\Rightarrow$$

$$x\frac{nml-ml-nl-mn}{nml}=a\Rightarrow x=\frac{anml}{nml-ml-nl-mn}.$$

The problem has a solution if $nml > n + m + l$, or, after dividing both sides by $lnm \neq 0, \frac{1}{n}+\frac{1}{m}+\frac{1}{l}<1$. Indeed, the portion cannot be greater than the whole.

The Closest Capitals. Let x mi be the length of the bridge, then

$$x=1+\frac{1}{5}x+\frac{1}{4}(x-\frac{1}{5}x-1)+\frac{3}{4}\Rightarrow x-\frac{1}{5}x-\frac{1}{4}x+\frac{1}{4\cdot 5}x=1-\frac{1}{4}+\frac{3}{4}$$

$$\frac{3}{5}x=\frac{3}{2}\Rightarrow x=\frac{5}{2}=2.5.$$

A 2.5-mile road-and-rail bridge between Kinshasa and Brazzaville could make them even closer.

Exceptionality of Capitals. Let x mi be the distance, then

$$x=a+\frac{1}{n}x+\frac{1}{m}(x-\frac{1}{n}x-a)+b\Rightarrow x-\frac{1}{n}x-\frac{1}{m}x+\frac{1}{mn}x=a-\frac{1}{m}a+b\Rightarrow$$

$$\frac{nm-m-n+1}{nm}x=\frac{bm+a(m-1)}{m}\Rightarrow x=\frac{(bm+a(m-1))n}{nm-m-n+1}.$$

Capitals Named after Presidents, Kings, and Tribes. Let x mi be the distance, then

$$x=779+\frac{1}{2}x+\frac{1}{3}x+\frac{1}{4}(x-779-\frac{1}{2}x-\frac{1}{3}x)$$

$$\Rightarrow x-\frac{1}{2}x-\frac{1}{3}x-\frac{1}{4}x+\frac{1}{4\cdot 2}x+\frac{1}{4\cdot 3}x=779-\frac{1}{4}\cdot 779$$

$$\frac{3}{24}x=\frac{3}{4}\cdot 779\Rightarrow x=4674.$$

Let us find the length of each segment: The first segment is given, 779 miles, the second segment is 4674/2 = 2337 miles, the third segment is 4674/3 = 1558, the fourth segment is 4674 – 779 – 2337 – 1558 = 0, that contradicts the problem that the four segments are of nonzero length.

This problem does not have a solution in reality. Indeed, Washington DC and Monrovia are not connected by land road, though the straight or flight distance is 4674 miles.

Multilingual Capitals. Let x mi be the distance, then

$$x = a + \frac{1}{n}x + \frac{1}{m}(x - \frac{1}{n}x - a) \Rightarrow x - \frac{1}{n}x - \frac{1}{m}x + \frac{1}{mn}x = a - \frac{1}{m}a$$

$$\Rightarrow x \cdot \frac{mn - m - n + 1}{mn} = a \cdot \frac{m-1}{m}$$

$$x \cdot \frac{(m-1)(n-1)}{mn} = a \cdot \frac{m-1}{m} \Rightarrow x \cdot \frac{(n-1)}{n} = a.$$

If $a = \frac{x(n-1)}{n}$, then $\frac{1}{m} \cdot (x - \frac{1}{n}x - a) = \frac{1}{m} \cdot (x - \frac{1}{n}x - \frac{x(n-1)}{n}) = 0$ for any m. Namely, the third portion of the distance is 0 that contradicts the problem condition of three nonzero parts.

The problem does not have a solution.

Sources

en.wikipedia.org/wiki/Mozambique

www.mapquest.com/travel/interesting-facts-about-madagascar/

www.google.com/search?q=distance

dynamictours.hu/10-facts-about-hungary-you-might-not-have-known-before/

worldstrides.com/blog/2017/10/14-interesting-facts-romania/

www.liquisearch.com/capital_city/distances_between_capital_cities_nearest_and_farthest

www.quora.com/Among-countries-that-share-borders-which-are-the-two-with-capitals-farthest-from-each-other

en.wikipedia.org/wiki/List_of_national_capital_city_name_etymologies

traveltriviachallenge.com/fun-facts-about-world-capitals/

www.cntraveler.com/story/the-worlds-closest-capitals-are-just-2-miles-apart

classroom.synonym.com/four-cities-named-after-presidents-7890674.html

traveltriviachallenge.com/fun-facts-about-world-capitals/

3.5 Mathematics of Games

Slinkies

The Slinky was accidently created by American naval engineer Richard James in 1943 when he was working on springs that could keep sensitive ship equipment steady at sea… The toy was a hit during its demonstration two years later when all 400 hand-made Slinkies were sold at $1 in 2 hours. Its simplicity and applicability have led to its manufacturing success. Over 500 million of them have been sold by now. In addition to playing with it, a Slinky has been used as a teaching tool, a holder, a portable, an extendable radio antenna, and so on. This popular toy has become a basis for Slinky Dog, Slinky Walking Spring Toy, and many other models. Slinky was inducted into the National Toy Hall of Fame in Rochester, New York, in 2000.

Problem. The length L of an idealized Slinky extended under its own weight is $L = \frac{W}{2k}$, where $W = mg$ is the weight of the Sinky, k is the spring constant, m is the Slinky's mass, and $g = 9.8\ \text{m/s}^2$ is the gravitation constant. The oscillation period T of a Slinky is $T = 2\pi\sqrt{\frac{m}{k}}$. Express dependence of the oscillation period on the Slinky's length.

Answer. $2\pi\sqrt{\frac{2L}{g}}$

Hopscotch

Kids all over the globe play hopscotch, though call it different names, *marelle* in France, *rayuela* or *golosa* in Spain, *stapu* or *kit kit* in India, *pada* in Zimbabwe, *seksek* in Turkey, *klassiki* in Russia, *tengteng* in Malaysia, *tiao fangzi* in China. Indeed, hopscotch is a simple game that does not require any equipment, just chalk or a stick to draw a court on the ground or pavement and a peg, a flat stone, or a flat tin to be tossed into numbered triangles or squares. Designs vary, but, traditionally, the court ends with a home base in which the player turns and completes the reverse trip. Hopscotch is so popular across continents that even the Guinness Book of World Records mentions the fastest hopscotch game.

Hopscotch came to us from prehistoric times. Perhaps kids played it in 1200 BC, though the first records in the Western world appeared in the 17th century.

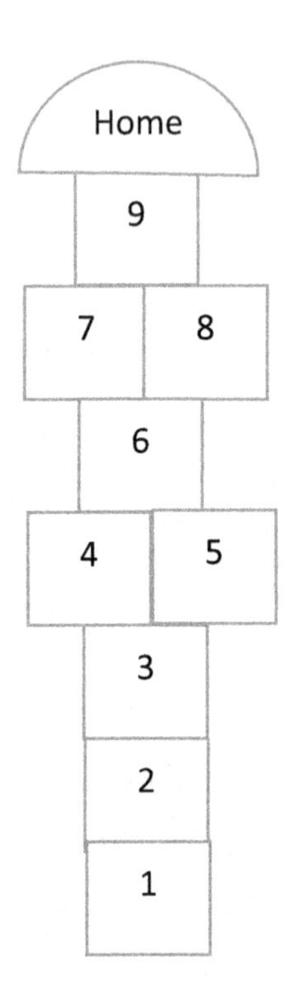

FIGURE 3.2

Problem. Determine the area and perimeter of the smallest rectangle that enclose a hopscotch court (see Figure 3.2) if the side of a square is 20 inches, and a stroke or the line separating squares is 2 inches wide.

A. 224 inches, 8188 sq. inches
B. 400 inches, 6400 sq. inches
C. 408 inches, 7268 sq. inches
D. 444 inches, 8096 sq. inches
E. 448 inches, 8188 sq. inches

Answer. E

Rummikub

Rummikub is a tile-based game where players arrange the tiles in groups following simple rules. One of the most interesting features of the game is that players can move and reuse the tiles already placed on the table. The game combines luck and strategy and changes quickly after each move. The winner is the first player who discards all their tiles.

FIGURE 3.3

Rummikub was invented by Romanian-Palestinian Ephraim Hertzano. The game has taken the path from being sold door-to-door to becoming a bestselling game. There are different variations of the game. Rummikub Tournaments and Championships are held worldwide.

Problem. The ratio between two Rummikub tiles put horizontally to three tiles put vertically (see Figure 3.2) is 20 to 21, but the difference is 0.4 inch. Find the dimensions of one Rummikub tile.

Answer. 2.8 inches by 4 inches

Ball Games

Spherical objects have always been fascinating not only to mankind, but to animals as well. Ball games came to us from prehistoric times. Their origins can be traced several millenniums back to the first civilizations of Mesoamerica, Maya, and Aztec. Ancient 3000-year-old balls made of hair-filled leather have been discovered in China. The oldest American ball found in Mexico is over 3000 years old. Ball games were depicted on Egyptian monuments and described in Greek plays. Our ancestors made a ball from animal bladders or skins and stuffed it with various materials. They used balls in their rituals, ceremonies, magic, forecasts, or games.

Originating in 2300 BCE the Chinese game Cuju aimed to keep the ball made of leather and stuffed with feathers off the ground by kicking it. FIFA has recognized Cuju as the ancient origin of football. The 12th century hollow ball made of deerskin was used for Japanese Kemari that aims to volley the ball among trees and keep it in the air for as many kicks as possible. It is said that the losing team in the Mayan's ritual game Pok-ta-Pok lost their heads. The Spanish were the first Europeans to see the bouncing rubber balls after Columbus returned from his voyages. In medieval times the British threw a ball from yard to yard to kick winter out and celebrate the new beginning of spring.

Football, baseball, basketball, bowling, golf, tennis, polo, and volleyball are played with a spherical ball though of different size, while American, Canadian, and Australian football are among those sports that use a spheroid ball.

Problem. Using the hints below, find the dimensions of football and basketball fields.

A. The area of a football field is 17 times the area of a basketball field.
B. The width of a football field is 40 meters more than the length of a basketball field.
C. The length of a football field is 7 times the width of a basketball field.
D. The length of a football field is divisible by 5 and 3.
E. All values that represent the length and width of a field are the smallest integers that satisfy the conditions above.

Answer. 105 m by 68 m, 28 m by 15 m

Darts

Darts games are popular target games played by throwing sharp objects or darts at a board with numbered spaces that determine the scores. Darts and boards have progressively changed over hundreds of years. The modern-day boards are usually circular, and the darts are feathered. Darting was invented in England in the 14th century, though its origin goes back much further. Darting used to be a military pastime and a training game for English archers and then widespread to other continents. Rules for the game have been developed, and clubs and competitions have been established over time. Now darts players from over 50 countries are members of the World Darts Federation and participate in the WDF World Cup, the Embassy World Professional Darts, and other championships.

Problem. A dart board is a regular octagon divided into nine segments by two vertical and two horizontal lines parallel to two sides, top, and bottom of the octagon (see Figure 3.4). A dart thrown at the board is equally likely to land at any spot on the board. Assuming that a dart hits the board, how likely does the dart land within the center square?

Answer. $\frac{\sqrt{2}-1}{2}$

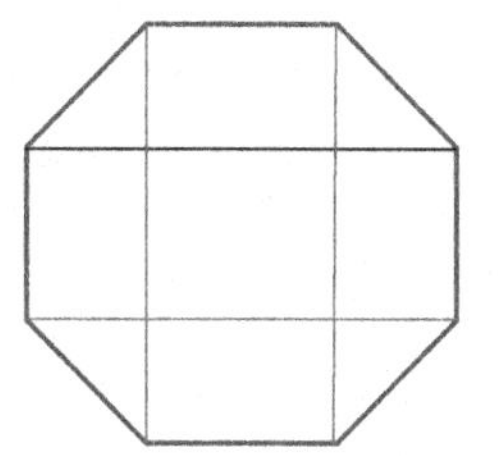

FIGURE 3.4

Dice Game

A die is a small polyhedron with marks on its sides used in playing games and gambling. The most common die is a cube with its six sides marked with one to six dots. The number of dots on opposite sides add up to seven. However, different numbers of sides, spots, and face designs are widespread.

Dice made of bone, ivory, bronze, agate, marble, amber, and other materials are the oldest gaming equipment known to humans for at least 5000 years. Cubic dice from 2000 BCE have been found in Egyptian tombs. Pyramidal or four-sided dice were found in Ur, which is in modern-day Iraq. After being thrown, fair dice lead to a random outcome and follow mathematical odds. In contrast, crooked or gaffed dice provide a desirable outcome. A cube slightly shaved down on one side will tend to settle down on its larger surface, a cube with a slightly convex side will roll off its convex sides to favor a cheater. Crooked dice have been found in the tombs of ancient Egypt, in Viking graves, and in prehistoric graves of North and South America.

Problem. Two fair six-sided dice are rolled. What is the probability that the number of dots that come up on the first die is not less than the number of dots that appear on the second die?

A. 1/3
B. 1/6
C. 1/2
D. 7/12
E. 2/3
F. 3/4
G. 5/6
H. 11/12
I. 35/36

Answer. D

Card Games

Playing cards were invented in China in the first millennium. Merchants brought them to Italy and Spain from the Islamic Mameluke empire of Egypt in the 14th century. Cards quickly became widespread throughout other European countries. The first playing cards were hand painted. The 15th century German invention of wood-block printing and French simplified design significantly reduced their cost, making cards more accessible to public. Numbered or illustrated playing cards have been used not only for playing games but also for education, fortune-telling, illusion, magic, tricks, and other purposes.

The most common deck consists of 52 cards, divided into four suits, each containing 13 ranks. The development of suits and ranks reflects the fascinating history and culture of the region and epoch, because each country made them applicable to it. The Mameluke suits were goblets, coins, swords, and polo-sticks. Polo was changed into clubs by Europeans. Germans suggested acorns, leaves, hearts, and bells, which French modified to trèfle, pique, hearts, and carreau before English modified them further. The rank of cards has also been altered over time. The oldest court cards were all males. The Germans were the first ones to introduce a queen but with the upper male superiority.

Problem. Eric and his date, Erica, have invited Victor with his date, Victoria, to play cards. The friends are seated at random around a square table, one person to a side. What is the possibility that Erica and Eric are seated opposite each other?

A. 1/2
B. 1/3
C. 1/4
D. 2/3
E. 3/4

Answer. B

Colored Balls

Selecting a certain item from an urn is an interesting and very common game. Although the result seems to be random, its prediction has a strong mathematical foundation.

Problem. Two-digit numbers of red and black balls are put into a sack. The average number of red and black balls is 90. How many balls of each color should be in the sack to have the largest probability of selecting a red ball?

Answer. 99, 81

Domino

Domino is a widespread tile-base game with gaming rectangular tiles called dominoes or bones. A line divides each domino in squares marked with dots or kept blank. A domino deck or pack contains one unique piece for all possible combinations of two ends. The number of dots in the most popular European domino, a so-called a double-six domino set, is between blank, one and six. In larger domino decks, like double-nine, double-12, double-15, and double-18, the maximum number of dots is a multiple of three and Arabic numerals are used for dots. Domino Whist, Matador, Muggins, Texas 42, Chicken Foot, Concentration, Double Fives, Threes, and Mexican Train are among many different versions of the game.

This popular game has a long history. Domino was mentioned in Chinese texts of the 12th century. Its modern version emerged in Italy in the 18th century. The game spread quickly to other European and American countries.

Problem. Each tile of a domino set is uniquely divided into two ends that are marked with dots or kept blank. Domino sets are distinguished by the maximum number of dots. How many tiles does a double-six domino set have? Develop a general formula for a different maximum number of dots on domino tiles. Apply this formula to a double-eight domino set suggested in Austria and popular double-nine and double-twelve domino sets.

Answer. 28, 45, 55, 91

Triomino

A triomino is a tile-base game like domino but with triangular tiles. A French immigrant to the United States, Allan Cowan, was captivated with mathematics of numbers and invented a triomino when he was just a teenager. Spending hours and years testing and improving his invention, he decreased the numbers on tiles from 0–9 to 0–5. Although numerous game publishers rejected his game, Mr. Cowan did not give up his teenage dream and finally the Pressman Toy Corporation accepted the game when he was in his late 30s. After 10 years, a Dutch game company introduced triominos to Europe. The game became so popular that it was awarded *the game of the year* in the Netherlands. Interestingly, France, which did not accept the triomino initially, is now one of its strongest markets.

Problem. Each side of a triomino triangular tile has a unique combination of three numbers from 0 to 5 arranged in a nondecreasing order clockwise. A repetition of numbers is allowed. Namely, combinations 1-4-5 and 1-4-4 work but 1-4-3 does not. How many tiles are in the standard triomino set with numbers from 0 to 5? Find a general formula to determine the number of tiles in any version of triomino sets. How many tiles are in the original (numbers 0 to 9) version of triomino set?

Answer. 56, 220

Solutions

Slinkies. The spring constant k can be found from $L = \frac{W}{2k}$ and $W = mg$ as $k = \frac{mg}{2L}$. Substituting the last expression to $T = 2\pi\sqrt{\frac{m}{k}}$ leads to

$$T = 2\pi\sqrt{\frac{2Lm}{mg}} = 2\pi\sqrt{\frac{2L}{g}}.$$

Hopscotch. The smallest rectangle is a rectangle that touches the base, top, and side squares. Its length is composed of 7 squares and one semicircle, connected by 9 strokes. The side of a square is 20 inches and the radius of a semicircle is also 20 inches, then the length of the rectangle is 8·20 + 9·2 = 178 inches. The width of the rectangle is determined by 2 squares and 3 strokes and is equal to 2·20 + 3·2 = 46 inches.

The perimeter of the rectangle is (178 + 46) ·2 = 448 inches and its area is 178·46 = 8188 sq. inches.

Rummikub. Let w be the width and l be the length, then

$$\frac{2l}{3w} = \frac{20}{21} \Rightarrow \frac{l}{w} = \frac{10}{7} \Rightarrow w = 0.7l$$

that together with $3w - 2l = 0.4$ gives $l = 4$, $w = 2.8$.

2nd way. Let us consider the system of equations in two variables

$$\begin{cases} \frac{2l}{3w} = \frac{20}{21} \\ 3w - 2l = 0.4 \end{cases} \Rightarrow l = 4, w = 2.8.$$

The Rummikub tile is 2.8 inches by 4 inches.

Ball Games. Let a and b be the length and width of a football field and c and d be the length and width of a basketball field, then from Hints A and C

$$\begin{cases} \dfrac{ab}{cd} = 17 \\ \dfrac{a}{d} = 7 \end{cases} \text{ follows } a = 7d \text{ and } \frac{b}{c} = \frac{17}{7} \text{ or } 7b = 17c.$$

From the relation $a = 7d$, the length of a football field a should be divisible by 7. Following Hint D, it should also be divisible by 5 and 3. Then $a = 7 \cdot 5 \cdot 3 = 105$ is the smallest integer to satisfy Hint E. Then $d = 15$. Combining the relation $7b = 17c$ and Hint B, we get $b = 68$, $c = 28$.

A football field is 105 m by 68 m and a basketball field is 28 m by 15 m.

Darts. The regular octagon is divided into nine regions, central square with the side a, four right triangles with sides a, $a/\sqrt{2}$, and $a/\sqrt{2}$, and four rectangles a by $a/\sqrt{2}$. The total area of the octagon is

$$a^2 + 4 \cdot \frac{a^2}{2 \cdot \sqrt{2} \cdot \sqrt{2}} + 4 \cdot \frac{a^2}{\sqrt{2}} = \frac{2a^2(\sqrt{2}+2)}{\sqrt{2}}.$$

The area of the central square a^2 takes

$$\frac{a^2\sqrt{2}}{2a^2(\sqrt{2}+2)} = \frac{a^2\sqrt{2}(\sqrt{2}-2)}{2a^2(\sqrt{2}+2)(\sqrt{2}-2)} = \frac{\sqrt{2}-1}{2}$$

portion of the regular octagon.

The probability that the dart lands within the center square is $\dfrac{\sqrt{2}-1}{2}$.

Dice Game. If two fair six-sided dice are rolled, then the number of possible outcomes is $6 \cdot 6 = 36$. If six dots come up on the first die, then any number out of 6 appeared on the second die provides a favorite outcome because it will be less than or equal to 6. Analogously, five dots on the first die gives five favorite outcomes because only six dots on the second die is greater than 5. Continue our discussion this way, we can finish with one dot on the first die producing one favorite outcome. Thus, the total number of favorite outcomes is the sum of numbers between 1 and 6, which is 21, that leads to the probability of favorite outcomes $21/36 = 7/12$.

The probability that the number of dots that come up on the first die is not less than the number of dots appeared on the second die is $7/12$.

Card Games. Only one of three friends, Erica, Victor, or Victoria, can be seated opposite Eric, which leads to 1/3 chance for Erica to be seated opposite her date.

The probability that Erica and Eric are seated opposite each other is 1/3.

Colored Balls. Let x and y be the number of red and black balls. Then $(x + y)/2 = 180$, or $y = 180 - x$. In order to have the largest number of red balls $\frac{x}{y} = \frac{x}{180 - x}$ should be maximum, that can be reached if the numerator takes the largest possible value, while the denominator takes the lowest possible value. Then x can be the largest two-digit number of 99 and the corresponding y is 81. Then the probability of selecting a red ball is $99/180 = 11/20$, and a black ball is $81/180 = 9/20$.

Actually, the largest two-digit number could be chosen at once.

There are 99 red balls and 81 black balls.

Domino. One domino tile is defined by numbers of dots on its ends independently of what end is considered first. Namely, there is one domino tile that can be read as a tile with two dots on the left end and three dots on the right end (2-3 tile) or with three dots on the left end and two dots on the right end (3-2 tile).

A double-six domino set has seven different options to fill in each end of a domino tile because the blank or a domino tile without any dots at the end is included in addition to one to six dots. So, only seven different options for dots can follow a domino tile with 0 (blank) dots on another end. Then, only six different options for dots, from 1 to 6, are left for a domino tile with one dot on another end because a 1-0 tile, which is also a 0-1 tile, has been considered previously. Following this algorithm, only five different options for dots, from 2 to 6, are left for the second end of a domino tile with two dots on another end, because 2-0 and 2-1 tiles, which are also 0-2 and 1-2 tiles, have been considered previously, and so on.

1st way. The results above can be presented in a table

Number of dots on one end of a domino tile	Different options of number of dots the second end avoiding repletion of tiles
0	7
1	6
2	5
3	4
4	3
5	2
6	1

Adding the numbers of the second column, 7 + 6 + 5 + 4 + 3 + 2 + 1 = 28, we get the number of tiles in the double-six domino set.

2nd way. This approach is based on the observation that the numbers in the second column are seven terms of an arithmetic sequence with the first term of 7 and difference of –1, or they can be also considered as an arithmetic sequence with the first term of 1 and difference of 1.

Using the formula for the sum of seven terms of an arithmetic sequence $S_7 = \frac{a_7 + a_1}{2} \cdot 7$, we get $S_7 = \frac{7+1}{2} \cdot 7 = 28$, that gives us the number of tiles in a double-six domino set.

This formula can be generalized for a domino set with the maximum number n of dots:

$$S = \frac{n+1+1}{2} \cdot (n+1) = \frac{n+2}{2} \cdot (n+1) = \frac{n^2+3n+2}{2},$$

where $n + 1$ instead of n is taken to consider tiles with no dots (blank) on its side. Substituting 7 for our double-six domino set, we get 28.

3rd way. This method considers the number of combinations of choosing m out of n items if order does not matter, defined by the formula

$$C(n, m) = \frac{n!}{m!(n-m)!}.$$

If $m = 2$, and $n + 1$ taken instead of n (as before), we get

$$C(n+1, 2) + (n+1) = \frac{(n+1)!}{2!(n+1-2)!} + (n+1) = \frac{n(n+1)}{2} + (n+1) = \frac{n^2+3n+2}{2},$$

which coincides with the formula above. The term $(n + 1)$ is added because of domino doubles, like 3 dots and 3 dots on both ends.

The general formula can be also presented as

$$S = \frac{n^2+3n+2}{2}.$$

For a double-eight domino set:

$$S = \frac{8+2}{2} \cdot (8+1) = 45.$$

For a double nine domino set:

$$S = \frac{9+2}{2} \cdot (9+1) = 55.$$

For a double-12 set:

$$S = \frac{12+2}{2} \cdot (12+1) = 91.$$

There are 28 tiles in double-six, 45 tiles in double-eight, 55 tiles in double-nine, and 91 tiles in double-12 domino sets.

Triomino. Let n denote any number between 0 and 5. The number n can be followed by any number m between n and 5, that in turn can be followed by any number between m and 5. For instance, if 3 is the first number, then the second number can be 3, 4, or 5 (three options). If 4 is selected as the second number, then the third number can be 4 or 5 (two options), and so on.

1st way. Keeping in mind that counting starts from 0 gives 6 different options to mark the next side of a triangle, let us prepare a table to visualize the process:

The 1st number	**The 2nd number** 0 1 2 3 4 5	
0	6 5 4 3 2 1	
1	5 4 3 2 1	
2	4 3 2 1	**Number of**
3	3 2 1	**options**
4	2 1	**for the 3rd**
5	1	**number**

Adding all numbers of options for the 3rd number we get that there are 56 triomino triangular tiles.

2nd way. A more mathematical way can be obtained if we notice that all numbers in each row provide an arithmetic sequence with the first term of 1 and difference of 1, and the sum of its terms of

$$S = \frac{a_i + a_1}{2} \cdot i \text{ or } S = \frac{i+1+1}{2} \cdot (i+1) = \frac{i+2}{2} \cdot (i+1),$$

where i is the number of terms in sequences that decreases from 6 to 1. Hence, the total sum of all rows is

$$S = \sum_{i=1}^{n} \frac{1+i}{2} i,$$

that can be simplified as

$$S = \frac{1}{2}\sum_{i=1}^{n} i + \frac{1}{2}\sum_{i=1}^{n} i^2 = \frac{1}{2}\cdot\frac{n+1}{2}n + \frac{1}{2}\cdot\frac{n(n+1)(2n+1)}{6}$$
$$= \frac{n(n+1)}{4}\left(1+\frac{2n+1}{3}\right) = \frac{n(n+1)(n+2)}{6}.$$

The standard triomino set includes 5 as a maximum end number, then $n = 6$ (numbers of 0 to 5), and the number of tiles is $6\cdot(6+1)(6+2)/6 = 56$.

The original triomino set has numbers of 0 to 9. Then $n = 10$, and the number of tiles is $10\cdot(10+1)(10+2)/6 = 220$.

The standard triomino set with the maximum possible end number of 5 has 56 tiles. The initially constricted triomino set with the maximum of 9 has 220 tiles.

Sources

en.wikipedia.org/wiki/Dominoes

rummikub.com/

en.wikipedia.org/wiki/Rummikub

en.wikipedia.org/wiki/Hopscotch

sites.psu.edu/ballgamesoftheworld/ancient-ball-games/

en.wikipedia.org/wiki/Ball

en.wikipedia.org/wiki/Slinky

www.theguardian.com/notesandqueries/query/0,5753,-2647,00.html

www.britannica.com/topic/playing-card

www.britannica.com/topic/dice

dartsguide.net/guides/dart-games-rules/

darthelp.com/articles/the-history-of-darts/

www.britannica.com/topic/darts

owlworksllc.com/inventor-stories/tenacious-tri-ominos-inventor-allan-cowan/;

en.wikipedia.org/wiki/Triominoes

3.6 Miscellaneous Mixtures

Olympic Games

The Olympic Games in ancient Greece were vital and passionate sporting, social, and cultural events for twelve centuries, though only males could attend and compete. The games were so popular that Greeks could not put their army together to defend against their enemies, like during the Persian invasion in 480 BCE, because many citizens attended the Olympic Games that year. For the first three centuries the games were held in the sanctuary of Olympia and lasted five days.

Luckily, the games are back after centuries of abandonment. Gold, silver, and bronze Olympic medals are awarded to the winners now, though only silver, copper, or bronze medals and an olive or laurel branch were given during the first modern Olympic Game in Athens, Greece, in 1896.

Problem. Olympic gold medals are made up of 92.5% silver and 7.5% gold and then plated with 6 grams of pure gold. To make Olympic gold medals a company uses two alloys. One alloy is 96% silver and 4% gold. Another alloy is 88% silver and 12% gold. How much of each alloy should be taken to make one gold medal that weighs 556 grams?

Answer. 312.75 g, 243.25 g

Olympic Medals

The Olympic Games originated in Greece almost three millennia ago, became very important by the end of the 6th century BCE, lost their popularity after the Roman victory over Greece in the 2nd century BCE, and were abandoned by Romans in the fourth century due to their pagan links. The victory wreaths were made from branches of olive trees.

When the Olympic Games came back, they brought new traditions, rules, and awards. The gold medals were made of solid gold for just a few years until the 1912 Stockholm Olympics. The host city designs and casts the Olympic medals.

Problem. The company has received the order to make Olympic gold medals with 92.512% silver and then plate them with pure gold. The alloy contains 371.8 grams more silver than copper, zinc, and other mixture metals. To get the content required to cast a gold medal, the company combines the alloy with the mixture of metals 7 times the weight of the mixture in the original alloy. What is the content of the original alloy? What is the weight of the gold that plated the medal if its weight is related to the weight of the silver as 3 to 203? How much does golden medal weigh? Round your result to the nearest tenth.

Answer. 99%, 6 g, 412 g

Saltiest Lakes

Can you imagine your life without salt? Your meal will not be that tasty. Being a source of sodium and chloride ions, salt plays an essential role in maintaining human health and regulating fluids in the body and blood pressure. Sea and lake water is its main source. With an average salinity level of 3.4%–3.6%, the salinity level of the oceans is much lower than the salinity level of some lakes. The saltiest water in the world with a salinity of 43% is Gaet'ale Pond located above a hot spring in Ethiopia. The pink waters of Lake Retba in Senegal and never frozen Don Juan Pond in Antarctica also have a very high salt content. A saline level of 35% in shallow lagoon Garabogazkol in Turkmenistan is higher than in the famous Dead Sea in Israel and Jordan. Lake Assal, a crater lake in the Danakil Desert in Djibouti, is among the saltiest lakes as well.

Problem. The Gaet'ale Pond has a salinity of 43%. The salt content is 40% in Lake Retba, 35% in Lake Assal, and 23% in the Dead Sea. Samples of water from these world's saltiest lakes have been taken to compare their water substances with a mixture from other lake samples. What samples from the mentioned above lakes and how much of them should be added to a 6-liter sample from the Gaet'ale Pond sample to obtain the minimum weight of solution with the same salinity level as Lake Assal has? Find the weight of the resulting solution.

Answer. 4l, 0l, 10l

Acids

An acid has particles that create hydrogen and can give its hydrogen ions to another substance. Acids have different pH levels with pH = 0 as the most acidic. The word *acid* means *sour* in Latin. Indeed, sour milk, vinegar, lemon juice, and other acids taste sour. Although the word *acid* is associated with something dangerous, acids bring many beneficial uses in our day-to-day lives.

Problem. The first jar contains five liters of a 54% acetic acid solution, the second jar holds a 48% acid solution. Two jars are mixed to get a 50% acid solution. How many liters are in the second jar?

Answer. 10 l

Acids Everywhere

Numerous types of acids are used in batteries as electrolytes, in fertilizer industries, in the pickling process to clean rust and other contamination from metal surfaces, in petroleum refining, in refrigerators, in air conditioners, and in household cleaning items.

Problem. The first bottle contains five liters of a 46% acid solution, the second bottle contains four liters of a 44% acid solution. The third bottle is filled with one liter of a solution with unknown acid content. Half of the solution from the third bottle was mixed with the solution from the first one. The remainder of the solution from the third container was added to the second bottle. The solution in the first and second bottles became 50% acid. How acidic is the solution in the third bottle? Is all the information presented in the problem necessary to answer the question?

Answer. 90%, no

Acids in Medicine

Various forms of acids are widely used in medicine, pharmacology, and nutritional science. Well-known aspirin, vitamin C, and salicylic acid are acids. Our body needs acids. Hydrochloric acid breaks down complex compounds into simpler ones that are easier to absorb. Our DNA contains acids in its structure. Amino acids are the basic unit of proteins in our bodies.

Problem. The first container contains five liters of a 48% acid solution, the second container contains six liters of a 45% acid solution. The third container has ten liters of acid solution. Part of the acid solution from the third container is added to the first container, and the remaining part is mixed with the second container to obtain a 50% acid solution in the first and second containers. What is the acid level in the solution in the third container? How much solution from the third container is added to the first and second containers?

Answer. 54%, 2.5 l, 7.5 l

Date Palms

Dates are high in fiber and antioxidants. Their nutritional benefits support brain health and prevent diseases. These oval-shaped reddish-yellow sweet fruits grow on the date palm trees that require arid conditions with long hot summers and little rainfall. Date palms were cultivated in the Middle East long before they were introduced to Southwest Asia, Northern

Africa, and Spain and then were brought to Mexico, California, and South America. Egypt is a world leader in date production and cultivation accounting for more than 17% of global date production. The fresh dates can be soft, semi-soft or dry, though the dry fresh date is not the same as the dried date.

Problem. Fresh dates contain 60% moisture. A bulk of 100 kg of dates was left to dry. After some moisture evaporated, the bulk contained 20% moisture. What was its weight?

Answer. 50 kg

Mango Trees

Mango is a large tropical tree native to India, Bangladesh, and Pakistan that grows over 100 feet in height and 12 feet in diameter. Indians domesticated and cultivated it around 2000 BCE. The mango was introduced to East Asia centuries later, then to the Philippines, Africa, and Brazil in the 15th century, and to other tropical countries later. The mango fruits come in orange, red, green, and yellow. Their flesh, peel, and seeds are packed with nutrients and antioxidants, contain immune-boosting nutrients, support heart health, help prevent diabetes, have low calorie content, and have a lot of other health benefits. Global mango production exceeds 43 million tons with India and China as its main supplying countries.

Problem. Mango contains 60% less flesh than moisture. A bulk of mango was left to dry. After half of the moisture evaporated, the bulk became 200 kg lighter. What was the original weight of the bulk?

Answer. 500 kg

Grapes and Raisins

Grapes are juicy and do not contain any fat but are a natural source of antioxidants and vitamins, help the immune system, prevent cancer, lower blood pressure, protect against heart disease and diabetes, reduce high cholesterol, help maintain brain health, slow down aging, and have other numerous health benefits. Raisins are dry grapes. China, Italy, and the United States of America are the largest grape producers in the world.

Problem. A bulk of 1000 kg of grapes was left to dry to produce raisins. It became half of its initial weight when the amount of moisture in the bulk went down to 80%. How much moisture was in grapes?

Answer. 90%

Nuts Mixture

Almonds, cashews, hazelnuts, Brazil nuts, pecans, pistachios, walnuts, and other nuts have numerous impressive health benefits. They are loaded with antioxidants and are a great source of vitamin E, magnesium, selenium, and many other vital nutrients. Despite being high in fat and low in carbs, nuts have contributed to weight loss. Finally, nuts are a tasty and suitable snack that fits all kinds of diets.

It is interesting that peanuts are legumes, like peas and beans, though they are referred to as nuts due to their similar nutritional characteristics.

Problem. Three sacks of mixed nuts have a total weight of 300 lb. To make sacks of the same weight, 4/5 of 1/3 of the content of the first sack is moved to the second bulk and 1/5 of the 1/3 of the content of the first sack is moved to the third one. Because the weights of bulks were still uneven, 2/13 from of the content of the third bulk is moved to the first one. Finally, when 1 lb from the second bulk is moved to the third bulk and 13 lb from of the second bulk is moved to the first one, all bulks weight the same. What is the initial weight of each bulk?

Answer. 105 lb, 85 lb, 110 lb

Solutions

Olympic Games. Let x g be the amount of the first alloy, then $(556 - x)$ g is the weight of the second alloy. The required amount of silver is

$$.96x + .88(556 - x) = .925 \cdot 556,$$

or $0.08x = 25.02$, that leads to $x = 312.75$.
If the amount of gold is considered, then

$$.04x + .12(556 - x) = .075 \cdot 556$$

leads to the same result of $x = 312.75$.

312.75 gram of the first alloy and 243.25 gram of the second alloy are required to make one gold medal.

Olympic Medals. Let s g and c g be the amount of silver and mixture in the original alloy that a company was given, then $s - c = 371.8$ and the weight of the original alloy is $s + c = 371.8 + 2c$. If 7 times the weight of the mixture in the original alloy is added, then the weight of the alloy becomes $371.8 + 2c +$

$7c$ with required 92.512% silver. The amount of silver remains. That leads to the equation

$$0.92512(371.8 + 2c + 7c) = s \text{ or } 0.92512(371.8 + 9c) = 371.8 + c.$$

Solving the last equations, we get $c = 3.8$ and $s = 375.6$. The amount of silver and mixture in the original alloy is 375.6 g and 3.8 g, respectively.

The initial alloy consists of $\begin{matrix} 379.4 & 100 \\ 375.6 & x \end{matrix} \Rightarrow$ 99% of silver and 1% of mixture.

The weight of the obtained alloy $371.8 + 2c + 7c$ becomes 406 g. Then the golden plate is $406 \cdot 3/203 = 6$ grams.

The Olympic gold medal weighs 412 grams.

Saltiest Lakes. Since a 23% salinity level of the Dead Sea is the lowest compared to the salinity levels of Lake Assal and Lake Retba, then to minimize a weight of the resulting solution only a sample from this lake should be mixed with 6 liters of the Gaet'ale Pond sample.

Let x be the amount in liters of the sample from the Dead Sea. Then $0.23x$ is its salt content and $(6 + x)$ is the weight of the resulting solution after adding 6 liters of the Gaet'ale Pond sample with a salinity of 43%. Hence,

$$6 \cdot 0.43 + x \cdot .23 = (6 + x) \cdot 0.35,$$

or $6 \cdot .08 = x \cdot .12$, that leads to $x = 4$.

A total of 4 liters of a sample from the Dead Sea should be taken to get 10 liters of solution with the same salinity level as in Lake Assal.

Acids. Let l be the amount in liters of a solution in the second jar. If two jars are mixed, the resulting solution becomes $(5 + l)$ of a 50% acid solution, that is, $0.54 \cdot 5 + .48 \cdot l = 0.5 \cdot (5 + l)$ or $0.02l = 0.2$ that leads to $l = 10$. In the equation 54% is presented as 0.54.

10 liters of a 48% acid solution are mixed with five liters of a 54% acid solution to get a 50% acid solution.

Acids Everywhere. 1st way. If three bottles are mixed, the resulting 10-liter mixture becomes 50% acid, that is,

$$0.46 \cdot 5 + .45 \cdot 4 + 0.a \cdot 1 = 0.5 \cdot 10,$$

where 46% is presented as 0.46, a% as $0.a$. Therefore, $a = 90\%$.

2nd way. 1/2 of the solution from the third bottle added to the first bottle makes the (5 + 1/2)-liter 50% acid solution. The equation that describes the process becomes

$$0.46 \cdot 5 + 0.a \cdot 1/2 = 0.5 \cdot (5 + 1/2),$$

from which $a = 90\%$ follows.

The solution is 90% acid. Either the amount of acid in two bottles or the amount of the solution in the third bottle mixed with the solution in the first bottle is needed to solve the problem if the second way is used.

Acids in Medicine. Let x and n liters be the acid level in the third container and the amount of the solution from the third container added to the first one. Then $10 - n$ is the amount of the solution from the third container added to the second one. Using a mathematical formulation to describe resulting mixtures in the first and second containers, we get

$$0.48 \cdot 5 + nx = 0.5 \cdot (n + 5) \text{ and } 0.46 \cdot 6 + x \cdot (10 - n) = 0.5 \cdot (10 - n + 6).$$

Simplifying both equations and rearranging them we obtain

$$0.5n + 0.1 = nx \text{ and } 5.3 - 0.5n = x \cdot (10 - n).$$

From the problem

$$x \neq 0,\ 0.5 \cdot (16 - n) - 2.7 \neq 0, \text{ and } n \neq 10.$$

Then dividing the first equation by the second and cancelling x lead to the equation in one variable

$$\frac{0.5n + 0.1}{5.3 - 0.5n} = \frac{n}{10 - n} \Rightarrow 0.4n = 1 \Rightarrow n = 2.5.$$

Substituting $n = 2.5$ to the first or second equation and solving it for x, we get $x = 0.54$.

2.5 liters of a 54% acid solution from the third container is added to the first container and 7.5 liters of it to the second containers.

Date Palms. There is 60% moisture in dates. Hence, the bulk of dates weighing 100 kg contains just 40 kg of flesh. When dates contain 20% moisture, 40 kg of flesh remain and take 100% – 20% = 80%, and, thus,

$$\begin{matrix} 40\text{kg} & 80\% \\ x\text{kg} & 100\% \end{matrix} \Rightarrow x = \frac{40 \cdot 100}{80} = 50\text{kg}.$$

The bulk of dates becomes 50 kg.

Mango Trees. Since mango contains 60% less flesh than moisture, then it holds 20% of flesh and 80% of moisture. Let x kg be the original weight of the bulk of mango. Following the problem, we can summarize that

Weight	Flesh	Moisture
x kg	$0.2x$ kg	$0.8x$ kg
$(x - 200)$ kg	$0.2x$ kg	$0.4x$ kg

Hence, $(x - 200) = 0.2x + 0.4x$ leads to $x = 500$.
500 kg of mango was left to dry.

Grapes and Raisins. Let x be a moisture portion in grapes. Hence, there is $1000x$ kg of moisture in the bulk of 1000 kg, and $(1000 - 1000x)$ kg is the grape flesh. When the moisture level drops down to 80% or takes 0.8 of the bulk weight, the grape flesh keeps the same amount of $(1000 - 1000x)$ kg that now becomes 20% of $0.5{\cdot}1000$ kg bulk, and $1000 - 1000x = 0.2{\cdot}0.5{\cdot}1000$ or $x = .9$
Grapes contain 90% moisture.

Nuts Mixture. The total weight of the three bulks is 300 lb. Let x lb, y lb, and z lb be the initial weights of each bulk. After 4/5 of 1/3 of the content from the first bulk is moved to the second bulk and 1/5 of the 1/3 of the content from the first bulk is moved to the third one, the weight of the first bulk becomes $\frac{2}{3}x$ and the second and third bulks weight

$$y+\frac{4}{5}\cdot\frac{1}{3}x = y+\frac{4}{15}x \text{ and } z+\frac{1}{5}\cdot\frac{1}{3}x = z+\frac{1}{15}x.$$

The next transaction that moves 2/13 from the third bulk to the first one changes the weight of the first and third bulks to

$$\frac{2}{3}x+\frac{2}{13}(z+\frac{1}{15}x) = \frac{132}{195}x+\frac{2}{13}z \text{ and } \frac{11}{13}(z+\frac{1}{15}x).$$

The weight $y+\frac{4}{15}x$ of the of the second bulk remains the same,

Finally, when 1 lb from of the second bulk is moved to the third bulk and 12 lb from of the second bulk is moved to the first one, their weights $\frac{132}{195}x+\frac{2}{13}z+12$, $y+\frac{4}{15}x-1-12$, and $\frac{11}{13}(z+\frac{1}{15}x)+1$ become the same and equal 100 lb. The obtained information can be combined in the system of four equations in three variables

$$\begin{cases} \frac{132}{195}x+\frac{2}{13}z+12=100 \\ y+\frac{4}{15}x-13=100 \\ \frac{11}{13}(z+\frac{1}{15}x)+1=100 \\ x+y+z=300 \end{cases} \Rightarrow \begin{cases} \frac{66}{195}x+\frac{1}{13}z=44 \\ y+\frac{4}{15}x=113 \\ \frac{1}{195}x+\frac{1}{13}z=9 \\ x+y+z=300 \end{cases}.$$

Substituting the third equation from the first one we get $\frac{65}{195}x = 35$, from which $x = 105$ follows. Then $y = 85$ by the second equation, and $z = 110$.

The initial weights of the first, second, and third bulks are 105 lb, 85 lb, and 110 lb.

Note. The resulting system is a system of *four* linear equations in *three* (one less than the number of equations) variables. One equation has not been used while solving the original system of equations.

Sources

www.worldatlas.com/articles/saltiest-lakes-in-the-world.html

earthobservatory.nasa.gov/images/84955/saltiest-pond-on-earth

www.guinnessworldrecords.com/world-records/76147-saltiest-lake

www.worldatlas.com/articles/world-leading-countries-growing-fresh-dates.html

www.worldatlas.com/articles/the-top-mango-producing-countries-in-the-world.html

www.atlasbig.com/en-us/countries-grape-production

en.wikipedia.org/wiki/Olympic_medal; www.britannica.com/sports/Olympic-Games

www.olympic.com

dewwool.com/uses-of-acids/#:~:text

3.7 World Religions

The Big Five

People have always believed in miracles, holy deities, and spiritual divine. Based on their social and cultural environment, they have developed prayers and rituals and built places for worship that related them to the supernatural realm. People have shaped religions! Although written evidence has only been found only from the last five millenniums, religion had been around long before this time. Some ancient religions, spiritual practices emerged and faded or were lost over time, while others have survived to these days. Some religions can be traced to their origins, others have very limited references.

The followers of the five or the *Big Five* religions, Christianity, Islam, Hindu, Buddhism, and Judaism, account for over three-quarters of the world's population. The Big Five are the most recognized and influential religions, though Shinto, Jainism, Voodoo, Confucianism, Taoism, and other smaller religions also play an important role in the modern society.

Problem. It is estimated that 31.7% of people are either Christians or Jewish, while 54.7% of people are either Christians or Muslims. 22.1% of people are either Hindus or Buddhists, while there are 6.9% more Buddhists than Jewish. 23% of people either belong to other religions or do not have any. How many people are believers of each of the Big Five religions?

Answer. 31.5%, 23.2%, 15%, 7.1%, 0.2%

Hinduism

Hinduism is not only the world's oldest and third largest religion but also a combination of ancient traditions, beliefs, and philosophies. Hindus believe in *reincarnation* with a constant cycle of being born, living, and dying on the path to enlightenment, *karma* that determines the level of new reborn, and the *Vedas' guidance.* They state that the divine essence or *atman* stays within each person and all creation, and that God goes by many names and presents in infinite ways. Brahma is the supreme deity or God responsible for creating everything in the universe. Shiva and Vishnu are primary Hindu deities as well. There are many other deities or devis, demi-gods or devas and goddesses.

Rather than having one sacred book, Hinduism has the Vedas, Samhitas, Upanishads, Ramayana, Bhagavad Gita, and many other religious books. The oldest found scripture of Hinduism, *Rig Veda,* is about 3500 years old. The uncovered bull and cow motifs, sacred in Hinduism, date back to 7000 BCE when the Indus River area was inhabited.

Hinduism has neither records of its founder nor central doctrinal authority. It has several sects that centered on one or several gods and follow different philosophical rituals, meditative paths, traditions, and yoga practices. Most Hindus live in India and then in Nepal, Mauritius, Bali, and Indonesia.

It is interesting that Hinduism was not originated in India but rather was brought and adopted there.

Problem. The Hinduism has $\sqrt[3]{xy+xz+yz}$ main denominations, where x, y, and z are positive integers that satisfy $\sqrt{x+y+z}=5$ and $x^2+y^2+z^2=497$. How many dominant sects are in Hinduism? Name those that you know.

Answer. 4

Abrahamic Religions

Christianity, Islam, and Judaism are among the five main religions having over a half of the world's population as followers of one of them. They are monotheistic religions worshiping only one God. Commandments or *Pillars* play an important role in their teachings providing religious guidelines. Each of these powerful religions has multiple denominations, such as Catholic, Orthodox, and Protestant in Christianity, Sunni and Shia in Islam, Orthodox, Conservative, and Reform in Judaism.

Christianity, Islam, and Judaism are *Abrahamic religions* in the sense that they claim to be direct or spiritual descents from Prophet Abraham, Jews as direct descents from Abraham's son Isaac, Muslims as direct descents from the second Abraham's son Ismael. Judaism was founded by Moses, although its history tracks back to Abraham who is considered to be the ancestor of all Jewish people. Judaism is the oldest out of these three religions. The six-pointed Star of David is its symbol. Christians consider themselves as spiritual through Jesus Christ's New Covenant.

Despite their common roots, the wars and violence among Christians, Muslims, and Jews have lasted for centuries.

Problem. Find the number $x + y$ of Commandments in Christianity and Judaism, and the number x of Pillars in Islam, if x and y are digits that satisfy the product

$$34x127\cdot321 = 11078y767.$$

List those that you know.

Answer. 10, 5

Buddhism

Unlike most ancient religions, Buddhism can be traced back to the teachings of one founder, prince Siddhartha Gautama, that gave up his wealth after witnessing the suffering outside the palace. He left his luxurious life and sat beneath the Bodhi tree or *the tree of awakening* until he reached enlightenment, thus, becoming *the Buddha* and referred to as the *Enlightened One*. There is a disagreement among scholars about the dates of the Buddha's birth and death. Contrasting with many other religions, the Buddha is not worshiped as a god. Buddhists focus on achieving *Nirvana*, or the state of enlightenment, where the believer will experience inner peace and wisdom. One of the most important Buddhist teachings is that life is full of suffering, which can be minimized by following the Noble Eightfold Path on the journey towards Nirvana.

Buddhists have many sacred writings that share the Buddha's philosophy and teachings, such as collections of Buddhist teachings in the Sutras and the Tipitaka. The three major branches of Buddhism, Theravada, Vajrayana, and Mahayana, have their own sacred texts and slightly varied interpretations of the Buddha's teachings, though all of them practice meditation, believe in reincarnation and rebirth, and consider *karma* that any deed will have a consequence later in life. Buddhism is reflected in the Triratna, that is, the *Three Jewels* of Buddha or (the teacher), *dharma* (the teaching), and *sangha* (the community).

Surprisingly, originating in India between the 6th and 4th centuries BCE, Buddhism did not stay there but spread to Central and Southeast Asia, China, Korea, and Japan where Buddhism has taken a central role in their spiritual, political, and social life.

Problem. The foundation of Buddhism is based on a set of x noble principles and y precepts or guidelines to which followers are expected to adhere. Find the number of noble principles and precepts if positive integers x and y satisfy

$$x^4 - y^2 = 231.$$

Answer. 4, 5

Holy Books

The printing press makes texts available in large quantities, unlike handwritten texts. The first printing press was created in China, though the date and inventors are unknown. The Diamond Sutra, which appeared during the Tang Dynasty, is the oldest known printed book. It is one of the most influential Buddhist texts.

Johannes Gutenberg from Germany invented the first printing press in the Western hemisphere. The Holly Bible was the first book printed by his press. Due to the fact that the texts of the Bible were previously copied by hand, there were several versions of the Bible. Not all Christian denominations used the same version that led to some disagreement among them. Gutenberg's invention contributed to this conflict even more causing, in some sense, the Protestant Reformation and appearance of multiple Christian denominations. Indeed, having the Bible available not only in their churches but also at home, Christians could read it, deeply think about its teachings, discuss them, and come to their own conclusions and visions creating new branches and sects of Christianity.

Problem. The Diamond Sutra was printed in the year of

$$\log_x^3 y - \log_x^2 y + \log_y^3 x - \log_y^2 x - 311.$$

The first Holy Bible was printed 587 years later. When were the Diamond Sutra and Bible available in print if $\log_x y + \log_y x = 11$?

Answer. 868, 1455

Twice-Holy Temples

The ancient city of Istanbul in Turkey, formerly known as Constantinople, has collected its treasures since it was founded as Byzantium by Greeks in the 7th century BCE. One of its most remarkable gems and the world's greatest monuments is Turkish Ayasofya, also called or used to be called Hagia Sophia, St. Sophia, Latin Sancta Sophia, Church of the Holy Wisdom, or Church of the Divine Wisdom. It was built as a Christian church in 538 and remained as such for almost millennium until the Turks took Constantinople after a long siege. Thousands of Christians took refuge in St. Sophia, despairingly waiting for a miracle to happen and save them... but it did not occur. The Turks turned the church into an Islamic Mosque, destroying lavish mosaics, covering up the symbols of the Christian religion, building four minarets, and carving inscriptions of Islam. Every Friday a priest reads the Quran there holding a drawn sword in his hand to show that this church was taken from the Christians by force. This historic site still reflects religious changes over time. Therefore, it became a museum in 1935 and then a mosque again in 2020.

History knows several examples when a holy temple dedicated to one religion was repurposed to serve a different religion. The largest temple in the world and one of the largest religious monuments, a Buddhist temple Angkor Wat, was built as a Hindu temple in honor of God Vishnu. The temple appears on the national flag of Cambodia.

Problem. What year did Turks turn Christian St. Sophia into an Islamic mosque if the remainder is 1 after dividing this year by 3, and the remainder is 10 if the year is divided by either 13 or 37?

Answer. 1453

Minarets

Islam is the second largest religion in the world. This monotheistic religion has begun in the 7th century BCE in the site where Saudi Arabia is now. The Quran is the holy book of Islam believed to be a revelation from the only one true God, Allah. A mosque is a place of prayer for Muslims, people that follow the Islamic religion. Mosques are constructed following the regional traditions and locations, but all of them have an open courtyard, an arcade, a prayer hall, and a minaret. Minarets are tall towers that aim to call for prayers and to emphasize the presence of Islam.

The difference between Sunni and Shia Muslims is in a disagreement over the succession to Muhammad. Sunni Islam believe that Muhammad did not name a successor and caliph is to be elected, while Shia Islam follows the tradition that Muhammad designated a successor.

Muslims live in many countries worldwide. Some countries follow Islamic law called *Shariat.*

Problem. The number of minarets that a mosque can have is described by the numbers of integers that the function

$$f(x) = \frac{14}{2+5e^{-x}}$$

can take. The maximum value that the function strives to reach but never reaches gives the number of world's countries with over 90% of Muslims, and $f(0)$ provides the number of main denominations of Islam.

How many denominations does Islam have? Can you name them? How many minarets does a mosque can have? In how many countries do Muslims make up 90% of the population? Can you name these countries?

Answer. 2, from 1 to 6, 7

Cathedrals

A Christian cathedral is the seat of a bishop and a center of worship. The word *cathedral* originated from Latin *cathedra* meaning *seat* that took this word from Greeks. The apse end of a Christian cathedral points east as being oriented towards Jerusalem, and the other three ends point west,

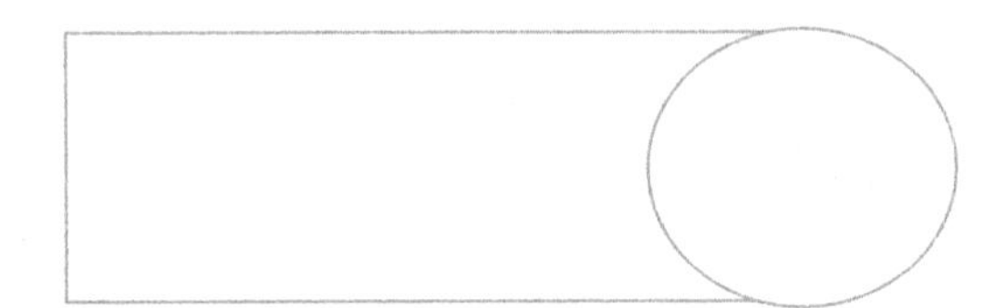

FIGURE 3.5

south, and north. It should be mentioned that cathedrals were often not exactly oriented east–west. Probably, the cathedral builders used the North Star as their orientation and approximately determined east. The magnetic compass was introduced in Europe only in the late 12th century, by the time that the sites of most cathedrals were well established. Moreover, many of them were built on sites of previous cathedrals, Roman and pagan temples and used their orientation.

La cathédrale de Notre Dame de Paris or *Our Lady of Paris* is one of the most well-known cathedrals. It is a medieval Catholic cathedral built on the island in the Siene River in the 12th century and dedicated to Virgin Mary. Because of its beauty and importance, the cathedral has appeared in numerous novels, movies, and documentaries.

Problem. La cathédrale de Notre Dame de Paris has an almost rectangular shape with a semicircular altar (see Figure 3.5). It is 128 m long and 48 m wide. What is the floor area of the cathedral?

A. 5897
B. 6144
C. 6802
D. 8611
E. 12 230
F. None of the above

Answer. A

Christian Churches

From the Middle Ages, Christian churches have been the tallest and most remarkable buildings in any place, city, or town. Some cities still do not allow any constructions higher than their major religion sites. The Ulm Minster is the main Lutheran congregation in Ulm, Germany. It is the tallest church in the world. The tallest Catholic and the tallest domed church building is the Basilica of Our Lady of Peace in Ivory Coast, though the tallest church building with two steeples and the tallest cathedral is Cologne Cathedral in Germany. The People's Salvation Cathedral when completed in Bucharest, Romania, will be the tallest Eastern Orthodox and the tallest domed cathedral.

The Sagrada Família in Barcelona, Spain, will be the world's tallest church when finished.

Problem. The Ulm Minster is 11 meters shorter than the Sagrada Família but 4 meters taller than the Cologne Cathedral. The People's Salvation Cathedral is the shortest among these four Christian temples that have the average height of 154.625 meters and the median of 159.5 meters. How tall are these Christian churches?

Answer. 161.5 m, 157.4 m, 127 m, 172.5 m

Statues of Buddha

Religions have influenced people for millenniums. They have painted sculptured images of their Gods and built places of worships. Each of them is marvelous in its own way.

Three of the four tallest statues in the world are of the Buddha. The Spring Temple Buddha was built in 2008 in Zhaocun, China. It is the second-tallest statue in the world. It includes the colossal statue of Vairocana Buddha that stands on a lotus throne atop an impressive pedestal. Vairocana Buddha is a major iconic figure in Buddhism and is seen as a universal Buddha and the illumination of wisdom. The Spring Temple Buddha takes its name from the nearby Tianrui hot spring famous for its healing properties. The construction uses 1100 pieces of copper cast, weights of 1000 tonnes, and costs $55 million.

Problem. The height of the statue of Vairocana Buddha takes 8/13 part of the entire height of the Spring Temple Buddha. The height of the lotus throne is 30 meters less than the height of the pedestal and their heights are related to as 5 to 11. How tall is the Spring Temple Buddha?

Answer. 208 m

Solutions

The Big Five. Using notations B for Buddhism, C for Christianity, I for Islam, J for Judaism, and H for Hindu and following the problem, we can write the system of five linear equations in five variables

$$\begin{cases} C+J=31.7 \\ C+I=54.7 \\ H+B=22.1 \\ B-J=6.9 \\ C+J+I+B+H=100-23 \end{cases}.$$

Substituting the first and the third equations to the fifth one, we get $31.7 + I + 22.1 = 77$, from which follows $I = 23.2$. Then $C = 31.5$ from the second equation and $J = 0.2$ from the first one. Finally, $B = 7.1$ and $H = 15$.

Christianity accounts for 31.5%, Islam 23.2%, Hindu 15%, Buddhism 7.1%, and Judaism accounts for 0.2% of the world's population.

Hinduism. Squaring both sides of $\sqrt{x+y+z} = 5$, we get $x+y+z = 25$. Squaring the last expression again and using $x^2 + y^2 + z^2 = 497$ lead to

$$25^2 = (x+y+z)^2 = x^2 + y^2 + z^2 + 2xy + 2xz + 2yz = 497 + 2(xy + xz + yz),$$

from which follows $xy + xz + yz = (625 - 497)/2 = 64$. Hence, $\sqrt[3]{xy + xz + yz} = 4$.

Shaivism, Shaktism, Vaishnavism, and Smartism are four main denominations of Hinduism.

Abrahamic Religions. The problem can be solved in various ways and using different multiplication methods, such as long multiplication method, grid multiplication method, and others.

Let us use the lattice diagram, also known as the Italian method, Chinese lattice, sieve multiplication, or Venetian squares, for multiplying $34x127$ by $321 = 11078y767$. The very last row below and the column to the left are added to record digits of the given products.

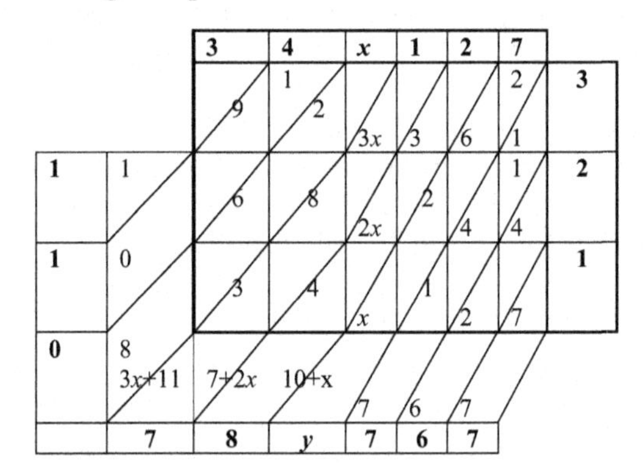

The product should be equal to $11078y767$ shown in the last row and the first column to the left. Indeed, the first 1 and the last three digits, 7, 6, and 7, coincide. The second digit from the left is 0 but it is 1 in the product that can be obtained if 2 is carried out from the previous sum and added to 8. Then $3x + 11$ is a two-digit number with 2 as tens.

The fourth digit from the end is $2 + 6 + 2 + x = 10 + x$ that should be equal to $10 + y$. Thus, $x = y$. The 1 is carrying to the next sum leading to $3x + 8 + 3 + 1 = 27$ (as earlier predicted, the first digit is 2) or $x = 5$.

Dedication to the faith, not making idols, not taking the name of the Lord your God in vain, honoring your father and mother, not murdering, not committing adultery, and not stealing, are among the Ten Commandments of Christianity and Judaism.

The *shahada* or the faith in Allah, *salat* or five daily prayers, *zakat* or helping the needy, *saum* or fasting during Ramadan, and *hajj* or pilgrimage to Mecca are the Five Pillars of Islam.

Buddhism. Since $x^4 - y^2 = 231$ can be rewritten as $(x^2 - y)(x^2 + y) = 231$ and 231 can be presented as the products of 1·231, 3·77, 7·33, 11·21, then for any positive integers we can get four systems

$$1.\begin{cases} x^2 - y = 1 \\ x^2 + y = 231 \end{cases} 2.\begin{cases} x^2 - y = 3 \\ x^2 + y = 77 \end{cases} 3.\begin{cases} x^2 - y = 7 \\ x^2 + y = 33 \end{cases} 4.\begin{cases} x^2 - y = 11 \\ x^2 + y = 21 \end{cases}.$$

The first, second, and third systems do not produce any integers as a solution. The fourth system leads to $2x^2 = 32$, or $x^2 = 16$. Then $x = 4$, $y = 5$.

The Buddha's original teachings are known as the *Four* Noble Truths that include *dukkha* (the truth of suffering), *samudaya* (the truth of the cause of suffering), *nirhodha* (the truth of the end of suffering), *magga* (the truth of the path that frees from suffering).

The five precepts dictate that Buddhists are against killing, stealing, lying, misusing sex, and intoxication.

Holy Books. Let us raise $\log_y x + \log_x y = 11$ to the second and third powers. Then, remembering $\log_x y = \dfrac{1}{\log_y x}$ that lead to $\log_x y \cdot \log_y x = 1$, we get

$$\begin{aligned} 11^2 &= (\log_x y + \log_y x)^2 = \log_x^2 y + 2\log_x y \cdot \log_y x + \log_y^2 x \\ &= \log_x^2 y + 2 + \log_y^2 x, \text{or} \log_x^2 y + \log_y^2 x = 119. \end{aligned}$$

$$11^3 = (\log_x y + \log_y x)^3 = \log_x^3 y + 3\log_x^2 y \cdot \log_y x + 3\log_x y \cdot \log_y^2 x + \log_y^3 x =$$

$$= \log_x^3 y + 3\log_x y + 3\log_y x + \log_y^3 x = \log_x^3 y + \log_y^3 x + 3 \cdot 11,$$

$$\text{or} \log_x^3 y + \log_y^3 x = 1298.$$

Then,

$$\begin{aligned} \log_x^3 y - \log_x^2 y + \log_y^3 x - \log_y^2 x - 311 &= (\log_x^3 y + \log_y^3 x) - (\log_x^2 y + \log_y^2 x) - 311 \\ &= 868. \end{aligned}$$

The Diamond Sutra, a Buddhist book, was printed in 868. The Christian Bible was printed in 1455.

Twice-Holy Temples. Since 3, 13, and 37 are co-primes and 10 gives the remainder 1 after division by 3, the number is $3 \cdot 13 \cdot 37 + 10 = 1453$.

The Turks turned Christian St. Sophia into an Islamic Mosque in 1453.

Minarets. Let us consider the behavior of the function $f(x)$

- $x \to \infty$, then $e^{-x} \to 0$, $2 + 5e^{-x} \to 2$, and $f(x) \to 7$ from below,
- $x \to -\infty$, then $e^{-x} \to \infty$, $2 + 5e^{-x} \to \infty$, and $f(x) \to 0$ from above,
- $x = 0$, then $f(0) = 14/(2+5 \cdot 1) = 2$.

Thus, the continuous function $f(x)$ takes integers from 1 to 6.

Sunni Islam and Shia Islam are the two denominations of Islam. Algeria, Bangladesh, Egypt, Iran, Iraq, Turkey, and Pakistan are seven countries where over 90% of their population are Muslims. Depending on the size, a mosque can have from one to six minarets.

Cathedrals. Since the width is 48 m, then the diameter of a semicircle is also 48 m, its radius is 24 m, and the length of a rectangular part is 128 – 24 = 108 m. Thus, the floor area is $104 \cdot 48 + \pi \cdot 24^2/2 = 5896.779$.

The floor area of Notre Dame is 5896.779 m^2.

Christian Churches. There are four (an even number) religious sites, thus the median is the average of the two churches with the middle height, which are Ulm Minster and Cologne Cathedral. Hence, their total height is $159.5 \cdot 2 = 319$ meters, and, considering the difference of 4 m between them, we can conclude that Ulm Minster is 161.5 meters tall, and the Cologne Cathedral is 157.5 meters tall. The Ulm Minster is 4 meters taller than the Cologne Cathedral, then its height is 161.5. Finally using the average height of 154.625 meters of the four churches, we can find that the People's Salvation Cathedral is $154.625 \cdot 4 - 161.5 - 157.5 - 161.5 - 11 = 127$.

The Ulm Minster is 161.5 m, the Cologne Cathedral is 157.4 m, the People's Salvation Cathedral will be 127 m, the Sagrada Família in Barcelona will be 172.5 meters.

Statues of Buddha. Let x m be the height of the Spring Temple Buddha statue, S, L, and P in meters be the heights of the statue of Vairocana Buddha, the lotus throne, and the pedestal. Then from $\frac{L}{P} = \frac{5}{11}$ and $L = P - 30$ follow

$$\frac{L}{L+30} = \frac{5}{11} \text{ or } 11L = 5(L + 30) \text{ that gives } L = 25,\ P = 55.$$

The total height of the Spring Temple Buddha statue is $x = S + L + P = 8/13 \cdot x + 25 + 55$ or $5/13 \cdot x = 80$, that leads to $x = 208$.

The Spring Temple Buddha is 208 m tall.

Sources

facts.net/religion-facts/

www.oldest.org/religion/religions

www.britannica.com/topic/Hagia-Sophia

en.wikipedia.org/wiki/Hagia_Sophia

facts.net/religion-facts/

whatdewhat.com/largest-religious-monuments-world/

life.org/hindu-denominations/

examples.yourdictionary.com/5-main-world-religions-and-their-basic-beliefs.html

www.khanacademy.org/humanities/ap-art-history/introduction-cultures-religions-apah/islam-apah/a/the-five-pillars-of-islam

www.newworldencyclopedia.org/entry/Abrahamic_religions

factcity.com/facts-about-buddhism/

www.history.com/topics/inventions/printing-press

www.archdaily.com/968046/the-architectural-cultural-and-religious-significance-of-minarets

www.khanacademy.org/humanities/ap-art-history/introduction-cultures-religions-apah/islam-apah/a/the-five-pillars-of-islam

en.wikipedia.org/wiki/Spring_Temple_Buddha

www.learnreligions.com/vairocana-buddha-450134

4

Discovering the World with Geometry

Experience proves that anyone who has studied geometry is infinitely quicker to grasp difficult subjects than one who has not.

~ Plato

There is no royal road to geometry.

~ Euclid

Captivating facts about legendary scholars, authors of famous books, and founders of the first academies and mathematical disciplines will be revealed through matching graphs with their equations offered in ten problems of the first section of this chapter. Their solution requires not only basic logical and analytic skills but also your imagination for graphing curves.

The problems from the next two sections uncover geometry in nature, architecture, and agriculture, as well as provide more astonishing information about the shapes of countries and symbols on national flags across the globe. Visualizing geometric contours, applying spatial reasoning, and linking shapes and solids enhance your understanding of geometrical fundamentals and assist you in recovering interesting details all over the world. Solve, discover, and enjoy!

4.1 From Graphs to Scientists

The First Female Scholar

She was born in 370 AD in Alexandria in Egypt that was under the Eastern Roman Empire at that time. Her father, the philosopher, Theon, aimed to raise a perfect human that could exceed in philosophy, sciences, medicine, and athletics. *She* significantly contributed to geometry and astronomy, taught philosophy, invented the hydroscope, and advocated universal education for women and children. *She* was also brave to dress like a scholar and drive her own chariot. *Her* life was brutally ended by a Christian mob who blamed her

DOI: 10.1201/9781003351702-4

TABLE 4.1

Equations of some common curves

Parametric equation	Cartesian equation
1. $y = \cos t,\ x = \dfrac{\sin t}{2}$	A. $x^2 + y^2 = 1$
	I. $x^2 y^2 + y^2 = 1$
2. $y = \cos 2t,\ x = \sin t$	O. $x^2 + y^2 = 2$
3. $y = \sec t,\ x = \tan t$	Y. $y = 1 - 2x^2$
4. $y = \sin t,\ x = \cos t$	H. $4x^2 + y^2 = 1$
5. $y = \sin t,\ x = \cos^2 t$	P. $y^2 - x^2 = 1$
6. $y = \cos t,\ x = \tan t$	T. $x = 1 - y^2$
7. $y = \cos 2t,\ x = \sin 2t$	

for religious turmoil, but her legacy lives on. *She* is the first notable female scientist and mathematician whose life has been relatively well recorded.

Problem. Name the first female mathematician if *her* name is decoded in Table 4.1, where the number of a curve parametric equation gives the position of a letter described by the corresponding Cartesian equation of the curve. Name the curves.

Describe similarities and differences between two equations: $y=\cos t$, $x=\sin t$ and $y=\cos 2t$, $x=\sin 2t$.

Answer. Hypatia

The Founder of the First Academy

This ancient Greek philosopher was born in Athens around 425 BCE. *He* founded the famous Academy in Athens in 587 BCE. The motto over the door of this philosophical school was *Let no one unversed in geometry enter here* because of *his* passionate belief that mathematics provided the finest training for mind and developed logical thinking. It is interesting that *he* is known by *his* pen name. His real name was Aristocles, but his wrestling coach nicknamed *him* due to his muscular build.

Problem. The name of one of the greatest and influential philosophers of Hellenic Greece is decoded in Figure 4.1 where a graph shows a position of the letter given by the corresponding equation. What is *his* name?

Answer. Plato

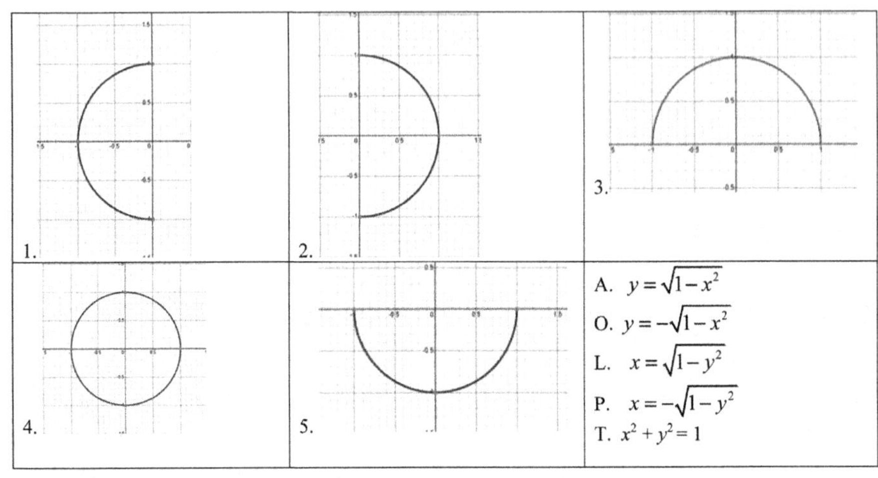

FIGURE 4.1

A Famous Brotherhood

Despite *his* incredible contribution to mathematics, music, and philosophy, very little is known about *his* life. *He* was born around 570 BCE in Greece and murdered in 495 BCE. *He* went to Egypt to study under temple priests and brought back one of the most important theorems named after *him*, through it was known to ancient Babylonians and Egyptians. *He* established a legendary school that discussed philosophy, art, and mathematics. This Brotherhood lasted over three centuries. It was so secret with strict rituals and prohibition to disseminate knowledge that it is still unknown whether statements named after *him* were solved by *him* or by his followers because all findings were named after *him*. *He* and his scholars discovered irrational numbers, found the sum of angles in a triangle, and started the modern number theory. They were the first to introduce rigorous mathematical proofs based on axioms and geometric ways of solving various problems. They discovered the intervals between musical notes and described the first four overtones building foundations of musical harmony. The brotherhood believed that *All Is Number*, meaning that everything in the universe depends on numbers.

Problem. Decode the name of the legendary Greek mathematicians and scientists. The letters of his name presented by graphs in Figure 4.2 are ordered following the numbers of their related equations. The fifth and ninth letters are the same.

1. $y = x + 1$
2. $y = x^2 + 1$

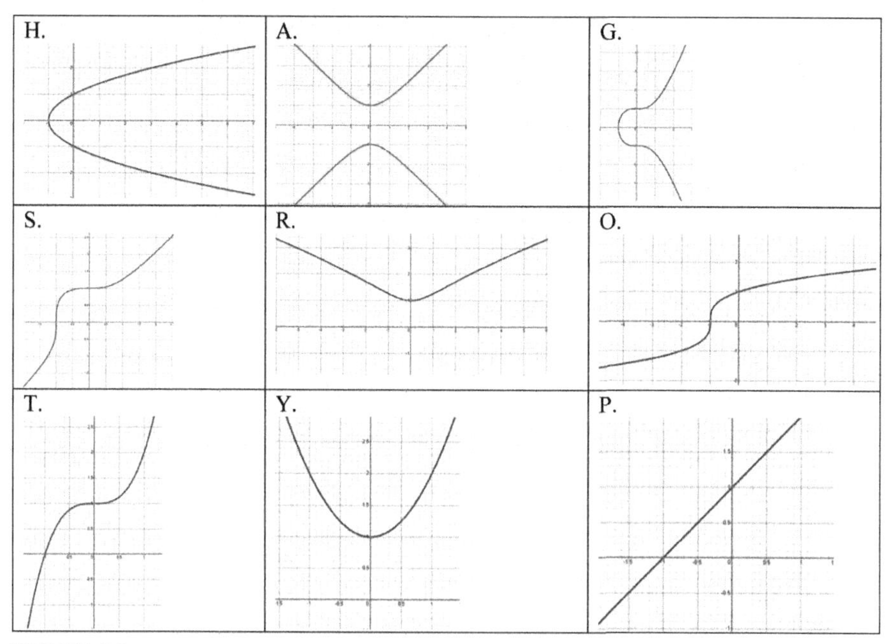

FIGURE 4.2

3. $y = x^3 + 1$
4. $y^2 = x + 1$
5. $y^2 = x^2 + 1$
6. $y^2 = x^3 + 1$
7. $y^3 = x + 1$
8. $y^3 = x^2 + 1$
10. $y^3 = x^3 + 1$

Answer. Pythagoras

The Designer of Paradoxes

This Greek philosopher of the fifth century BCE is the founder of dialectic. *He* is also famous for designing paradoxes that pushed the development of mathematics and logical thinking. In particular, *his* paradox *The Dichotomy* demonstrates that the motion can never even begin. *The Arrow* paradox shows that the arrow never moves. Attempts to explain these paradoxes led to the development of integral and differential calculus and only then were they resolved.

Problem. Answering the four questions below helps to find the name of the famous inventor of paradoxes. Just put all the letters side by side. Consider that the equation $Ax^2 + By^2 + Cx + Dy + E = 0$ is given.

1. If the graph of $Ax^2 + By^2 + Cx + Dy + E = 0$ is a circle and $A = 1/4$, then the value of B should be

S. –2 W. –4
T. –½ X. –¼
U. 2 Y. 4
V. ½ Z. ¼

2. The graph of the equation $Ax^2 + By^2 + Cx + Dy + E = 0$ with $A = 1$, $B = 1$, and nonnegative C, D, and E, is a circle if the following statement holds.

A. $E < 0$ E. $C^2 + D^2 > 4E$
B. $E > 0$ F. $C^2 + D^2 > E$
C. $C + D > E$ G. $C^2 + D^2 < 4E$
D. $C + D < E$ H. $C^2 + D^2 < E$

3. The graph of the equation $Ax^2 + By^2 + Cx + Dy + E = 0$ is a circle with the center in the origin (0,0), and nonnegative A and B, then the values of the coefficients are

I. $A = 1, B = 1$ M. $A = B, C = D$
J. $A = 0, B = 0, E > 0$ N. $A = B, C = 0, D = 0, E < 0$
K. $A = B, E > 0$ O. $A = B, C = 0, D = 0, E > 0$
L. $A = B, E < 0$ P. $A = B, C = 0, D = 0, E = 0$

4. The graph of $Ax^2 + By^2 + Cx + Dy + E = 0$ and $A = 1$, $B = 1$, is presented in Figure 4.3 with points (3, 3), (3, –5), (–1, –1), (7, –1) indicated in Figure 4.3. Then the values of other coefficients are

O. $C = -6, D = 2, E = -6$ T. $C = -3, D = 1, E = -5$
P. $C = 6, D = -2, E = 6$ U. $C = 3, D = -1, E = 5$
R. $C = 3, D = -1, E = 4$ V. $C = -1, D = 3, E = -6$
S. $C = -3, D = 1, E = -4$ W. $C = 1, D = -3, E = -6$

Answer. Zeno

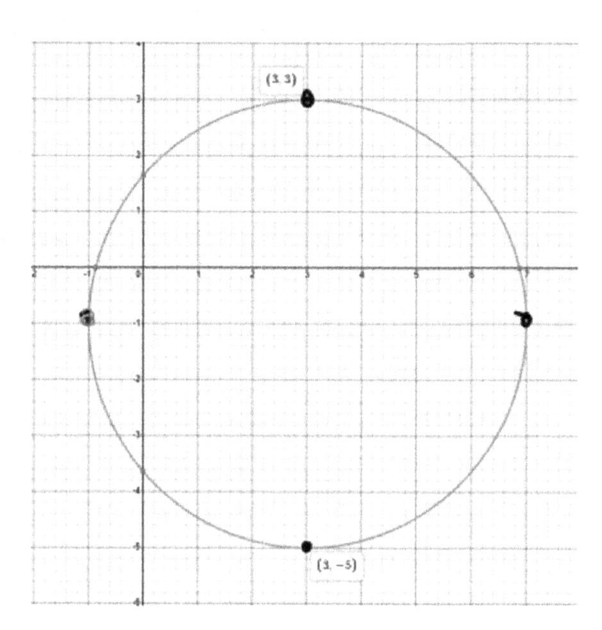

FIGURE 4.3

The Father of Geometry

He came from Greece to establish Mathematics School at the University of Alexandria. *He* is called the *Father of Geometry* because of his *Elements*, one of the most influential books. The mathematical treatise was written in Greek, then translated into Arabic, later into Latin, and now to other languages. The book is a collection of fundamentals of geometry, number theory and elementary algebra and contains five axioms and five postulates that we use today. Its chapters are studied at all schools in the world. Currently, there are three main geometries in the world, Riemann, Lobachevsky, and the third one named after *him*, which is the most common and traditional. Despite very little being known about his life, some of his books and wisdom sayings have survived to this day. It is said that answering Ruler Ptolemy's request for studying geometry fast, *he* replied *There is no royal road in Geometry.*

Problem. Decode the name of the famous Greek mathematician and geometer if the letters of his name presented by equations that describe the graphs below. The number of the graph presented in Figure 4.4 shows the position of a letter that describes the related equation.

L. $(y+1)^2+(x+2)^2=4$
C. $(y-1)^2+(x+2)^2=4$

U. $(y + 1)^2 + (x - 2)^2 = 4$
E. $(y - 1)^2 + (x - 2)^2 = 4$
D. $(y + 1)^2 + (x + 2)^2 = 2$
I. $(y - 1)^2 + (x - 2)^2 = 2$

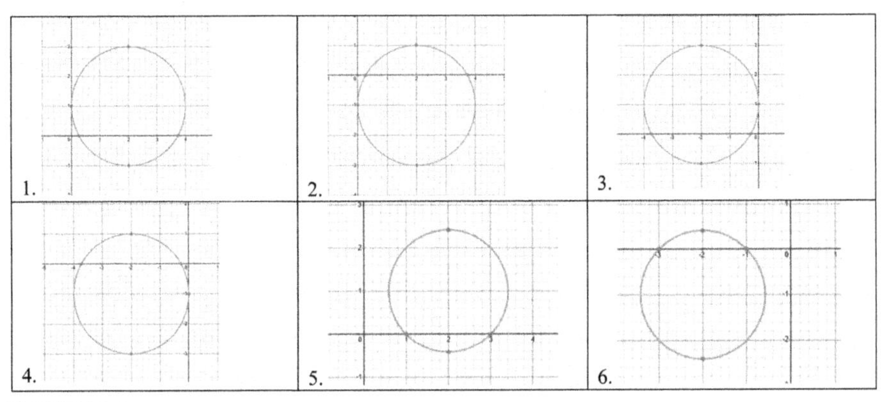

FIGURE 4.4

Answer. Euclid

Inventors of Logarithms

He was born in Scotland in 1550 when his father was just 16. Because of *his* political and religious views and anti-Catholic publications, *he* was often sent to his estate and forbidden to write in those areas. Thus, *his* creative mind concentrated on other issues. *He* predicted the future development of the machine gun, the submarine, and the army tank. *He* described the army tank as a chariot with a living mouth of mettle that would scatter destruction on all sides. *He* is best known for discovering logarithms and inventing a calculating devise named after *him*. *He* also made the use of the decimal point common. There are many fun stories about *him* and *his* behavior.

Problem. The name of a Scottish creator of logarithms is decoded in the following relation where the number of an equation shows the position of a letter given by the corresponding graph in Figure 4.5. Who invented logarithms?

1. $y = \log_2 x + 2$
2. $y = \log_2(x + 2)$
3. $y = \log_{1/2} x$

4. $y = \log_2 2^x$
5. $y = \log^2{}_2 x$
6. $y^2 = \log_2 x$

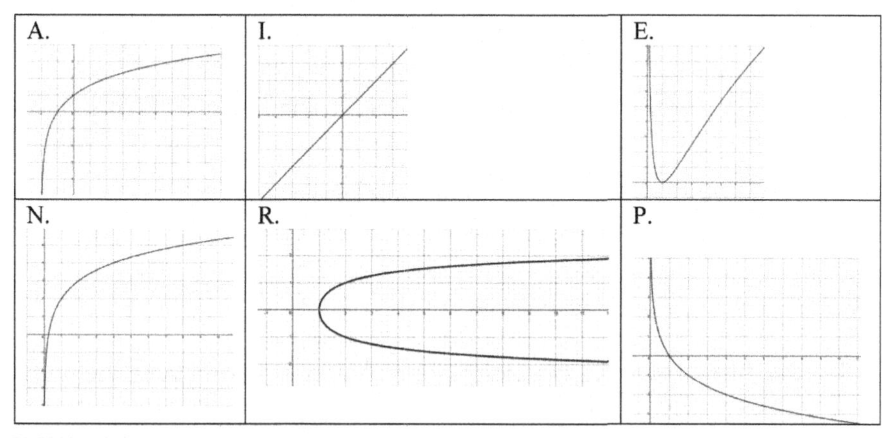

FIGURE 4.5

Answer. Napier

Creators of Analytic Geometry

Two great geniuses lived in the 17th century in France. Both were not mathematicians though their contribution to mathematics is vital. The first one was the father of modern philosophy famous for *his* statement *I think and therefore I am. He* wrote several books presenting his philosophical ideas. An appendix *La Geometrie* for one of his books delivered influential mathematical statements widely used today. Several mathematical rules presented there have been named after *him*. The second person worked for a local parliament and devoted his leisure time to mathematics enriching many of its branches. *He* is considered as a founder of the modern number theory and probability theory. Several mathematical statements have been named after *him*.

They independently invented analytic geometry offering different approaches. The first scientist considered a locus and then found its equation. *He* used modern notations of that time. The rectangular system of coordinates has been named after *his* Latinized name. The second inventor formulated the equation and then came up with its locus. He used archaic language to present his ideas.

Remark. The *locus* of points is the set of the points that satisfy some property.

Problem. The names of two French men who developed analytic geometry can be found if the following riddle is decoded.

- The word description of the locus of points, their graphs, and equations are given.
- The number of each word description of the locus shows the position of a letter given by the related letter bullet of a graph or an equation.
- The name of the first scientist consists of nine letters hidden in letter bullets of graphs. Its second and eight letters are the same. Its third and ninth letters are also the same.
- The name of the second scientist has six letters given by letter bullets of equations.

Word descriptions of the locus of points.

1. The distance from any point (x, y) on the curve to a single point $(c, 0)$ is constant and equals a. The graph and equation are given for $c = 1, a = 2$.
2. The sum of the distances from any point (x, y) on the curve to two foci $(-c, 0)$ and $(c, 0)$ is constant and equals $2a$. The graph and equation are shown for $c = 4$ and constant of 10.
3. The absolute value of the difference between the distances from any point (x, y) on the curve to two foci $(0, -c)$ and $(0, c)$ is a constant. The graph and equation are shown for $c = 5$ and constant of 4.
4. The product of the distances from any point (x, y) on the curve to two foci $(-c, 0)$ and $(c, 0)$ is constant and equal to 1. The graph and equation are presented for $c = 1$.
5. The product of the distances from any point (x, y) on the curve to two foci $(-c, 0)$ and $(c, 0)$ is constant and equal to 1/2. The graph and equation are presented for $c = 1$.
6. The product of the distances from any point (x, y) on the curve to two foci $(-c, 0)$ and $(c, 0)$ is constant and equal to 3/2. The graph and equation are presented for $c = 1$.
7. All points (x, y) on the curve are equidistant from the focus $(0, c)$ and line $y = c$, called the directrix. The graph and equation are shown for $c = 1$.

Graphs are shown in Figure 4.6.
Equations.

A. $(x^2 + y^2)^2 - 2(x^2 - y^2) = -0.75$

B. $y = \dfrac{x^2}{4}$

E. $\dfrac{y^2}{3^2} + \dfrac{x^2}{5^2} = 1$

F. $(x-1)^2 + y^2 = 2^2$

M. $(x^2 + y^2)^2 - 2(x^2 - y^2) = 0$

R. $\dfrac{y^2}{3^2} - \dfrac{x^2}{4^2} = 1$

T. $(x^2 + y^2)^2 - 2(x^2 - y^2) = 1.25$

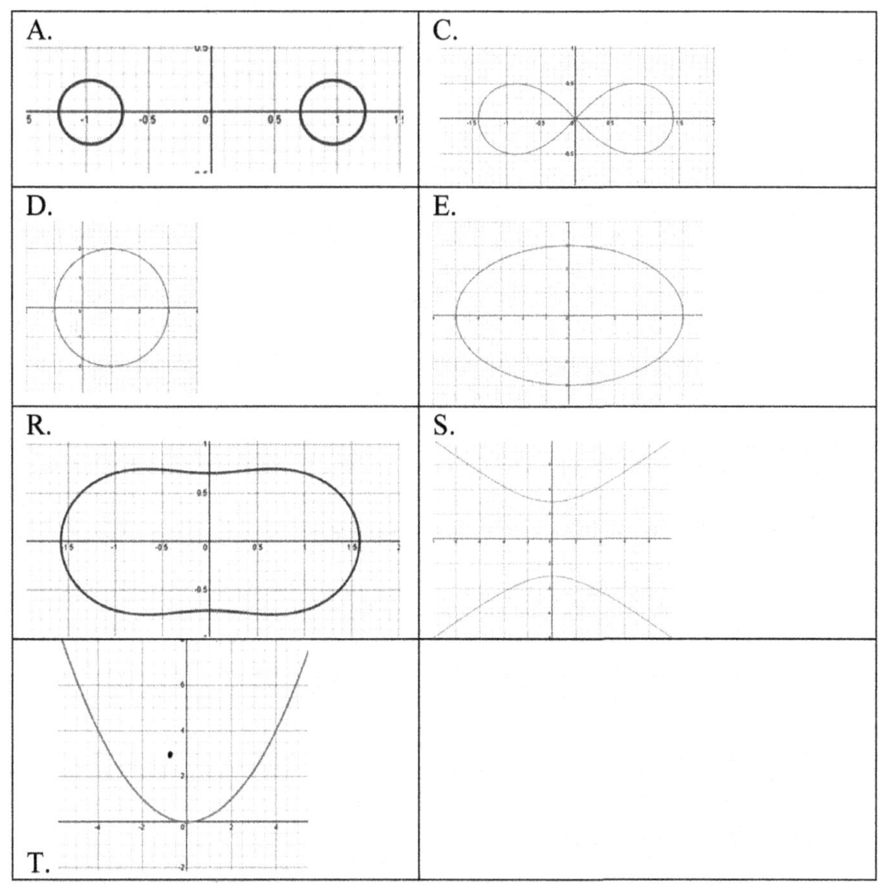

FIGURE 4.6

Answer. Descartes, Fermat

Scientists in the Names of Curves

Working as a German diplomat, *he* provided prominent contributions to different disciplines. *He* founded the Berlin Academy of Science in 1700. Being self-educated, *he* fluently spoke several languages, was charming, had many friends and admirers, though he never married. This universal genius wrote on philosophy, theology, ethics, politics, law, history, and philology. *He* improved the binary number system and created a cataloging system for librarians. Independently of Isaac Newton, *he* developed the differential and integral calculus based on a more understandable geometrical approach and suggested convenient notations of integrals and dx. *He* also suggested *coordinates, abscissa, ordinate,* and other mathematical terms. A basic rule of calculus, laws in philosophy, an asteroid, a lunar crater, university, library, and institutes have been named after *him*.

Problem. Decode the name of the German diplomat who was one of the calculus inventors, if the number of an equation shows a position of the letter related to the corresponding graph in Figure 4.7. Find applications of these curves in nature, science, art, architecture, and other disciplines.

1. *Folium of Descartes* $x^3 + y^3 - 3axy = 0$ was first proposed by French philosopher René Descartes in 1638 who also challenged Pierre de Fermat to find the tangent line to the curve. Fermat's solution to the problem contributed to the invention of calculus. The graph is shown for $a = 1$.
2. The areas between any two consecutive full turns around the *Fermat's spiral* $r^2 = a^2 \cdot \theta$ are equal. Mathematics was a hobby for French lawyer Pierre de Fermat. The graph is shown for $a = 1$.
3. *Witch of Agnesi* $y = \frac{8a^3}{x^2 + 4a^2}$ was studied by Fermat, Grandi, Newton, and Agnesi, though it was named after an 18th-century Italian mathematician and philosopher Maria Gaetana Agnesi who was the first woman to become a university professor of mathematics. Use $a = 1$ for the graph.
4. The curve *Cissoid of Diocles* $(x^2 + y^2)\, x = 2ay^2$ was named after Ancient Greek geometer Diocles who used it for solving a problem of doubling a cube using Euclidean tools. Take $a = 1$ for a graph.
5. A *Cassini oval* $(x^2 + y^2)^2 - 2a^2(x^2 - y^2) + a^4 = b^4$ was named after the Italian–French astronomer Giovanni Domenico Cassini who studied them in the late 17th century. He believed that the Sun traveled around the Earth on one of these ovals, with the Earth at a focus of the oval.

   Take $a = 1$, $b = \sqrt{\frac{3}{2}}$ for a graph.

6. The curves $y^n = k(a - x)^p \cdot x^m$ were studied by de Sluze in 17th century and were named *Pearls of Sluze* by Blaise Pascal. René François Walter de Sluze was born in Spanish Netherlands, now Belgium, and later studied civil and canon law, but enjoyed mathematics the most. The graph is for $a = 2$, $n = 2$, $k = 1$, $p = 3$, and $m = 4$.
7. The *Archimedean spiral* $r = a\theta$ was named after the 3rd century BCE Greek mathematician and engineer Archimedes. It describes the trajectory of a point moving away from a fixed point with a constant speed along a line rotated with constant angular velocity. The graph is for $a = 1$.

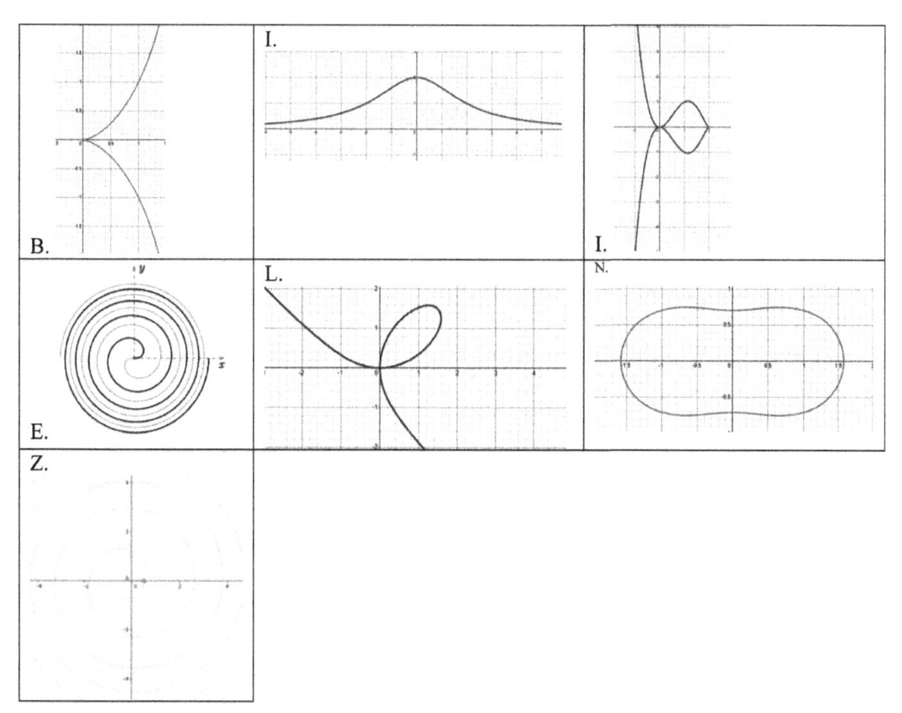

FIGURE 4.7

Answer. Leibniz

Authors of Famous Books

With their legendary collection of books, these two prominent scientists of antiquity opened new horizons in mathematics, though there is a lot of uncertainty about their lives.

Probably, the first scientist lived in the 3rd century AD in Alexandria, Egypt. In *his* most important series of books *Arithmetica, he* offered the analytical treatment of algebraic number theory, introduced many notations, and presented tricky mathematical problems with solutions. A new field of mathematics originated from this book. A branch of number theory and algebraic equations with integer coefficients and solutions have been named after *him*.

The second scientist was an Ancient Greek geometer and astronomer born about 262 BC in Southern Asia Minor. In his most notable collection of books called *Conic Sections, he* introduced the terms *ellipse, parabola,* and *hyperbola* and studied these curves. Only seven books have survived to our time.

Problem. Name authors of *Arithmetica* and *Conic Sections* if their ten-letter names are decoded by two sets of equations related to graphs shown in Figure 4.8. The number of the graph provides the position of the related letter in a name. The first set of equations contains the name of the author of *Arithmetica.* The second set of equations reveals the name of the author of *Conic Sections.*

Graphs.

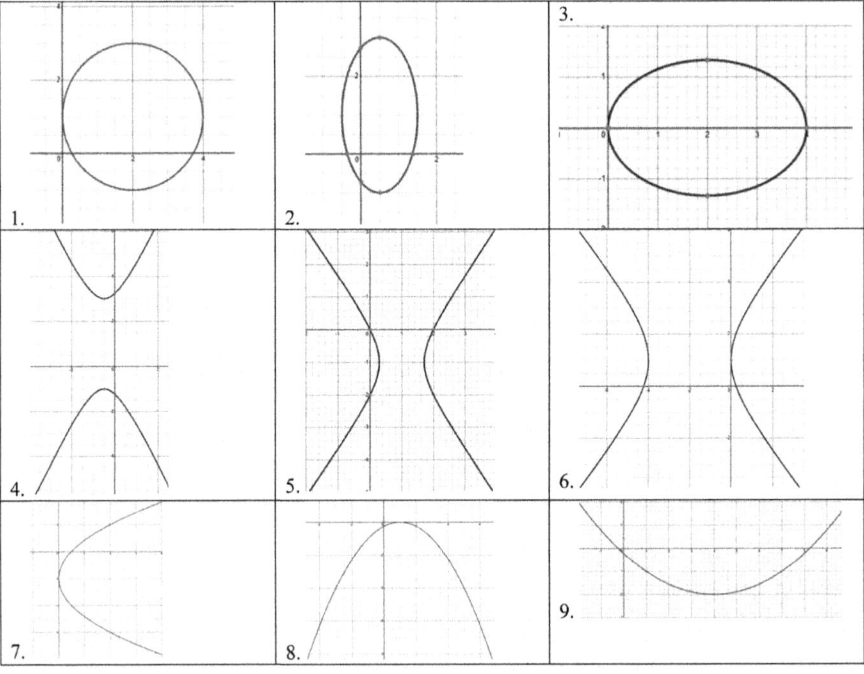

FIGURE 4.8

First set of equations.

A. $-y^2 + x^2 + 2y + 4x - 1 = 0$

I. $y^2 + 4x^2 - 2y - 4x - 2 = 0$

D. $y^2 + x^2 - 2y - 4x + 1 = 0$

U. $x^2 - 9y - 8x - 2 = 0$

H. $-y^2 + 2x^2 - 2y - 4x = 0$

N. $y^2 + 4y - 4x + 4 = 0$

O. $9y^2 + 4x^2 - 16x = 0$

P. $y^2 - 4x^2 - 2y - 4x - 4 = 0$

S. $y^2 - 4x^2 - 2y - 4x = 0$

T. $x^2 + 4y - 2x + 1 = 0$

Second set of equations.

L. $-\frac{(y+1)^2}{2} + (x-1)^2 = \frac{1}{2}$

L. $\frac{(y-1)^2}{2^2} - (x+1/2)^2 = 1$

O. $\frac{y^2}{2^2} + \frac{(x-2)^2}{3^2} = \frac{2^2}{3^2}$

O. $-(y-1)^2 + (x+2)^2 = 2^2$

A. $(y-1)^2 + (x-2)^2 = 4$

I. $y = -\frac{(x-1)^2}{2^2}$

N. $x = \frac{(y+2)^2}{2^2}$

P. $\frac{(y-1)^2}{2^2} + (x-1/2)^2 = 1$

S. $(y-1)^2 = (2x+1)^2$

U. $y = \frac{(x-4)^2}{3^2} - 2$

Answer. Diophantus, Apollonius

Creators of Notations

Although born in Basel, Switzerland, in the 18th century, *he* spent most of his life working in St. Petersburg Academy, Russia, pioneering influential discoveries in mathematics, physics, astronomy, geography, engineering, just to name a few disciplines. *He* introduced the summation sign Σ, symbols for a mathematical function $f(x)$, sides a, b, and c, the inradius r, and semiperimeter s of a triangle ΔABC, the exponential form of a complex number $re^{i\theta}$, and many other mathematical notations. One important and widely used constant and several mathematical statements have been named after *him*. During *his* prolific long life, *he* published over 700 books and papers. When *he* became partially and later totally blind, *he* made all complicated mathematical calculations in his mind and dictated *his* books and papers. Surprisingly, some incorrectness has been found only in three works *he* dictated.

Problem. Answer questions about graphs that depict the equation $Ax^2 + By^2 + Cx + Dy + E = 0$ and relations among its coefficients to define the name of the prominent Swiss scholar.

1. All coefficients in the equation $Ax^2 + By^2 + Cx + Dy + E = 0$ are nonzero. Then the only graph that is NOT possible is a graph of

   A. a circle
   B. an ellipse
   C. a hyperbola
   D. two lines
   E. a parabola

2. A graph of the equation $Ax^2 + By^2 + Cx + Dy + E = 0$ is given in Figure 4.9. What statement on the equation coefficients does describe the graph the best?

   O. $A = 0, B = 0, C > 0, D > 0, E < 0$
   P. $A = 0, B = 0, C > 0, D > 0, E > 0$
   R. $A > 0, B < 0, C > 0, D = 0, E < 0$
   S. $A > 0, B > 0, C > 0, D > 0, E < 0$
   T. $B > A > 0, C = 0, D = 0, E < 0$
   U. $A > B > 0, C = 0, D = 0, E < 0$
   V. $A > B > 0, C = 0, D = 0, E > 0$
   W. $A > B > 0, C = 0, D = 0, E > 0$

3. What changes in the coefficients of the equation $Ax^2 + By^2 + Cx + Dy + E = 0$ should appear to get the graph as in Figure 4.9?

   Note. Figure 4.9 contains two graphs, the original and the new one.

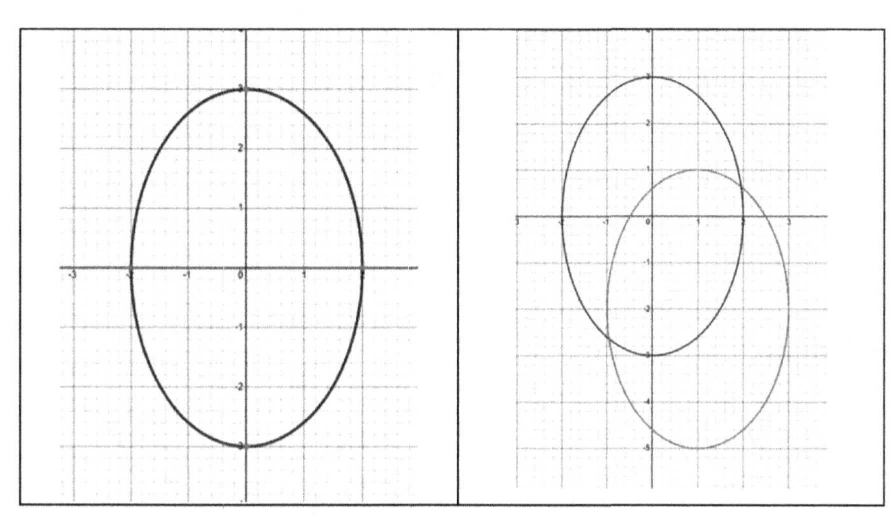

FIGURE 4.9

J. A and B are the same as in the original graph, $C > 0$, $D < 0$, E is decreased.

K. A and B are the same as in the original graph, $C > 0$, $D < 0$, $E > 0$ is increased.

L. A and B are the same as in the original graph, $C < 0$, $D > 0$, $E < 0$ is increased.

M. A is increased, B is decreased, $C < 0$, $D > 0$, $E < 0$ is increased.

N. A is decreased B is increased, $C < 0$, $D > 0$, $E < 0$ is decreased.

O. A is decreased B is increased, $C < 0$, $D > 0$, $E > 0$ is increased.

4. The sign of what coefficient in the equation $Ax^2 + By^2 + Cx + Dy + E = 0$ should be changed to get a graph of a hyperbola that opens up and down? The rest of coefficients remain the same.

   H. E
   I. D
   J. C
   K. B
   E. A

5. The value of what coefficient in the equation $Ax^2 + By^2 + Cx + Dy + E = 0$ should be modified to have a graph of a parabola that opens to the left or right? The rest of coefficients stay the same.

R. $A = 0$

S. $B = 0$

T. $C = 0$

U. $D = 0$

V. $E = 0$

Answer. Euler

Solutions

The First Female Scholar. Rewriting Equation 1, $y = \cos t, x = \frac{\sin t}{2}$, as $y = \cos t$, $2x = \sin t$ and applying the identity $\sin^2 t + \cos^2 t = 1$ of trigonometric functions, we get $4x^2 + y^2 = 1$, which is the Cartesian Equation H of an ellipse. Applying the same property to Equations 4 and 7, we get $x^2 + y^2 = 1$, which is the Cartesian Equation A of a circle. Equation 5 leads to Cartesian Equation T of the parabola $x = 1 - y^2$ that opens to the left. Remembering that $\cos 2t = \cos^2 t - \sin^2 t$. Equation 2 can be presented as $y = \cos 2t = \cos^2 t - \sin^2 t$, which, after adding $2\sin^2 t$ produces the Cartesian Equation Y of the parabola $y = 1 - 2x^2$ that opens down. The combination of trigonometric identities $\sec^2 t = 1 + \tan^2 t$ and $\sec t = 1/\cos t$ changes Equation 3 to the Equation P of a hyperbola and Equation 6 to Equation I.

Hypatia is the first female scholar and mathematician.

The Founder of the First Academy. The equation $x^2 + y^2 = 1$ of a circle tells us the fourth letter T. The square root by itself is nonnegative. Thus, y in $y = \sqrt{1 - x^2}$ and x in $x = \sqrt{1 - y^2}$ are positive and their graphs are above the x-axis and to the right of the y-axis respectfully. Hence, the third letter is A and the second letter is L. The square root with the negative sign has negative values. Thus, y in $y = -\sqrt{1 - x^2}$ and x in $x = -\sqrt{1 - y^2}$ are negative and their graphs are below the x-axis and to the left of the y-axis respectfully. Hence, the first letter is P and the fifth letter is O.

Plato founded the famous Academy in Athens in 587 BC where he taught the philosophical doctrines known as Platonism.

A Famous Brotherhood. A thoughtful analysis of equations and graphs leads to the correspondence between them. It is easy to see that that Equation 1 is the equation of a straight line with the y-intercept (0, 1), which is Graph P.

Equations 2 and 4, $y = x^2 + 1$ and $y^2 = x + 1$, are equations of parabolas $y = x^2$ and $x = y^2$ moved one unit up or left correspondingly, given by Graphs Y and H.

Graphs T and O are obtained by transforming basic graphs $y = x^3$ and $x = y^3$ one unit up or left, that is, they are described by Equations 3 and 7.

Equation 5, $y^2 = x^2 + 1$, is an equation of a hyperbola $y^2 - x^2 = 1$ presented as Graph A. Considering symmetry about the x-axis and y-axis and behavior of functions at infinity, correspondence between Equations 6 and 8 and Graphs G and R are visible.

Finally, the dynamics of x and y in Equation 10, $y^3 = x^3 + 1$, is like their dynamics in $y^3 = x^3$ or $y = x$ when both x and y are large, but different at their small values, which leads to Graph S.

Pythagoras of Samos established a legendary school to discuss philosophy, mathematics, music, and other disciplines.

The Designer of Paradoxes. Let us consider each case.

1. The coefficient of x^2 and y^2 should be the same in an equation that represents a circle. Thus, the correct answer is Z.
2. The equation $Ax^2 + By^2 + Cx + Dy + E = 0$ with $A = 1$ and $B = 1$, can be rewritten as $x^2 + y^2 + Cx + Dy + E = 0$, or $(x^2 + 2 \cdot C/2 \cdot x + (C/2)^2) - (C/2)^2 + (y^2 + 2 \cdot D/2 \cdot x + (D/2)^2) - (D/2)^2$ $+ E = 0$, that can be simplified to $(x + \frac{C}{2})^2 + (y + \frac{D}{2})^2 = \frac{C^2}{4} + \frac{D^2}{4} - E$. In order for a circle to exist, its radius should be positive, or $\frac{C^2}{4} + \frac{D^2}{4} - E$ > 0. Thus, the correct answer is E.
3. Since the graph is a circle, $A = B > 0$. The center of a circle is defined as $\left(-\frac{C}{2A}, -\frac{D}{2A}\right)$ then the values of the coefficients C and D should be zeroes for not moving the center of a circle from (0, 0). If moved to the right side, the coefficient -E represents the radius squared and should be positive. Thus, $E < 0$, and the correct answer is N, $C = 0$, $D = 0$, $E < 0$.
4. Since the distance between the point (3, 3) and (3, –5) and the points (–1, –1) and (7, –1) is 8, the radius is 4. Because the middle point between (3, 3) and (3, –5) is (3, –1) and between (–1, –1) and (7, –1) is (3, –1), the equation of the circle is $(x - 3)^2 + (y + 1)^2 = 4^2$ or $x^2 + y^2 - 6x + 4y - 6 = 0$, which is Option O.

Zeno of Elea (495 BC – 430 BC) was a Greek philosopher famous for his paradoxes.

The Father of Geometry. All equations represent a circle $(y + a)^2 + (x + b)^2 = r^2$ with the center at $(-b, -a)$ and the radius r.

Greek scholar Euclid, called the *father of geometry*, presented all advances in geometry and mathematics known at that time his book *Elements*.

Inventors of Logarithms. A brief analysis of equations leads to the correspondence between equations and graphs. In particular, Equation 4 can be simplified as $y = \log_2 2^x = x\log_2 2 = x$, that is $y = x$, which is the equation of a line graphing as Graph I. Only Graph E has nonnegative values of y, which can be produced only by Equation 5. Because of y^2, Equation 6 is symmetric about the y-axis, which leads to Graph R. It is obvious that Equation 3 and Graph P are related. Graphs A and N are obtained by moving the basic graph $y = \log_2 x$ 2 units left (Equation 2) or up (Equation 1) correspondingly.

John Napier discovered logarithms and invented Napier's bones.

Creators of Analytic Geometry. The decoding is based on the formula for the distance $d = \sqrt{(x_1 - x_2)^2 + (y_1 - y_2)^2}$ between any two points on a curve.

Let us consider Hint 1. Using the formula for the distance between any two points on a curve, we can write the distance, which is constant, between any point (x, y) and a point $(c, 0)$, as $\sqrt{(x-c)^2 + y^2} = a$. Squaring both sides of the statement, we get$(x^2 - c)^2 + y^2 = a^2$, which is the equation of a circle, and the focus $(c, 0)$ called the center. Thus, we get Graph D and Equation F.

Similarly, Hint 2 describes an ellipse, which is Graph E and Equation E, while Hint 3 describes a hyperbola, which is Graph S and Equation R.

Since Hints 4, 5, and 6 are similar, let us derive a general equation for these cases. Using the formula for the distance between any two points on a curve, we can write the product of the distances from any point (x, y) to two foci $(-c, 0)$ and $(c, 0)$, which is constant, as

$$\sqrt{(x-c)^2 + y^2} \cdot \sqrt{(x-(-c))^2 + y^2} = a.$$

Squaring both sides, we get $(x^2 - 2cx + c^2 + y^2)\cdot(x^2 + 2cx + c^2 + y^2) = a^2$, which after opening parentheses and rearranging terms becomes $(x^2 + y^2)^2 - 2c^2(x^2 - y^2) = a^2 - c^4$. If $c = 1$ and $a = 1$, then we get $(x^2 + y^2)^2 - 2(x^2 - y^2) = 0$, which is Equation A and Graph A. If $c = 1$ and $a = 1/2$, then we get $(x^2 + y^2)^2 - 2(x^2 - y^2) = -0.75$, which is Equation T and Graph R. If $c = 1$ and $a = 3/2$, then we get $(x^2 + y^2)^2 - 2\,(x^2 - y^2) = 1.25$, which is Graph T.

In Hint 7, all points on the curve are equidistant from the focus $(0, c)$, and directrix $y = -c$. Then

$$\sqrt{(x-x)^2 + (y-(-c))^2} = \sqrt{(x-0)^2 + (y-c)^2},$$

which after squaring both sides, becomes $y = \dfrac{x^2}{4c}$, which is the equation of a parabola that opens upward, which is Equation T and Graph B.

In summary, we obtained relations among Hints, Graphs and Letters as 1-D-F circle, 2-E-E an ellipse, 3-S-R a hyperbola, 4-C-M Lemniscate of Bernoulli, 5-A-A a Cassini oval, 6-R-T a Cassini oval, 7-T-B a parabola.

René Descartes (1596–1650) and Pierre de Fermat (1607–1665) independently developed analytic geometry using different approaches.

Scientists in the Names of Curves. A close look at curves and equations helps us to find some correspondence between them. For instance, y approaches $-\infty$ as x approaches ∞ and y approaches ∞ as x approaches $-\infty$ in the *Folium of Descartes* $x^3 + y^3 - 3axy = 0$ presented in Hint 1. The only curve with this property is Curve L. Thus the first letter is L. Then, the curve described by the *Witch of Agnesi* $y = \dfrac{8a^3}{x^2 + 4a^2}$ shows that y approaches 0 as x approaches $-\infty$ or $+\infty$. The only graph with this property is I, that gives the third letter.

The variable r is squared, $r^2 \geq 0$, in the polar equation of the *Fermat's spiral* $r^2 = a^2 \cdot \theta$, thus, there is a symmetry with respect to r, that leads to Graph E as the second letter.

The equation of a *Cassini oval* $(x^2 + y^2)^2 - 2a^2(x^2 - y^2) + a^4 = b^4$ remains the same if $-x$ or / and $-y$ are substituted for x or / and y. Hence, its graph is symmetric about the x-axis or / and y-axis, which leads to Graph N.

All curves are widely used in applications. There are just a few examples: Fermat's spirals appear in modelling electrical engineering processes, plant growth, and the shapes of spiral galaxies. The Witch of Agnesi is used in probability theory and approximation of functions by polynomials.

Prominent diplomat and scholar Gottfried Wilhelm von Leibniz (1646–1716) founded calculus and its notations, developed a water pump run by windmills, designed principles of binary codes, and made other contributions to politics, law, philosophy, theology, ethics, history, and philology.

Authors of Famous Books. There are different ways to solve the problem.

The 1st way is based on the formula $(a \pm b)^2 = a^2 \pm 2ab + b^2$ and rules of graph transformation. Indeed, let us take Equation I from the first set of equations, $y^2 + 4x^2 - 2y - 4x - 2 = 0$, rearrange terms, and complete a perfect square as $(y^2 - 2y + 1^2) - 1 + ((2x)^2 - 2 \cdot 2x \cdot 1 + 1^2) - 1 - 2 = 0$. The last equation becomes $(y - 1)^2 + (2x - 1)^2 = 4$ that after factoring 2 from $(2x - 1)^2$ and dividing both sides of the equation by 2^2, turns into $\dfrac{(y-1)^2}{2^2} + (x - 1/2)^2 = 1$, which is Equation P from the second group of equations. It is the equation of an ellipse that moved 1 unit up and ½ unit to the right, or Graph 2.

The 2nd way is to simplify equations from the second set. For instance, let us take Equation L from the second set of equations, $\dfrac{(y-1)^2}{2^2} - (x + 1/2)^2 = 1$. After opening parenthesis and multiplying both sides by 4, we obtain $y^2 - 4x^2 - 2y - 4x - 4 = 0$, with is Equation P from the first group of equations

Finding other letters is similar and leads to the relations 1-D-A, 2-I-P, 3-O-O, 4-P-L, 5-H-L, 6-A-O, 7-N-N, 8-T-I, 9-U-U, 10-S-S.

The 3rd way is to consider graphs first. For instance, Graph 1 is a circle with $r = (4 - 0)/2 = 2$ and center at (2,1), that leads to Equation D from the first set and Equation A from the second.

Diophantus of Alexandria is famous for his book *Arithmetica*. Apollonius of Perga is the author of *Conic Sections*.

Creators of Notations. Let us consider each hint.

1. An equation of a parabola contains either x^2 or y^2, but not both. Since all coefficients in the equation $Ax^2 + By^2 + Cx + Dy + E = 0$ are non-zero, a graph of a parabola is not possible, and the correct answer is E.
2. The axes of the ellipse intersect at the origin. Thus, $C = 0$, $D = 0$. The graph of the ellipse stretches more along the y-axis, then $A > B > 0$. The coefficient E should be less than 0 to be positive when moved to the right side. Therefore, the correct answer is U.
3. The ellipse is moved down ($D > 0$) and to the right ($C < 0$) but keeps its shape (A and B are the same). Due to changes in C and D, the negative $E < 0$ should be increased. The correct answer is L.
4. The sign of A should be negative to get a graph of a hyperbola that opens up and down, that leads to the correct answer of E.
5. The sign of A should be zero to have a graph of a parabola that opens to the left or right, that gives the correct answer of R.

The irrational constant 2.718281828459045… was named e after Swiss scholar Leonhard Euler (1707–1783).

Sources

en.wikipedia.org/wiki/Hypatia

en.wikipedia.org/wiki/Plato

www.desmos.com/calculator

en.wikipedia.org/wiki/Pythagoras

www.sutori.com/en/story/pythagoras-timeline--myg3wgdhKctRYMkFrSP3LRof

www.britannica.com/biography/Zeno-of-Elea

mathworld.wolfram.com/ZenosParadoxes.html

topfacts.org/11-interesting-facts-about-euclid/

en.wikipedia.org/wiki/John_Napier

en.wikipedia.org/wiki/Locus

en.wikipedia.org/wiki/Lemniscate

en.wikipedia.org/wiki/Cassini_oval,

en.wikipedia.org/wiki/Rene_Descartes
en.wikipedia.org/wiki/Pierre_de_Fermat
en.wikipedia.org/wiki/Folium_of_Descartes
en.wikipedia.org/wiki/Fermat%27s_spiral
en.wikipedia.org/wiki/Witch_of_Agnesi
en.wikipedia.org/wiki/Cissoid_of_Diocles
en.wikipedia.org/wiki/Gottfried_Wilhelm_Leibniz,
en.wikipedia.org/wiki/Cassini_oval
en.wikipedia.org/wiki/Diophantus
en.wikipedia.org/wiki/Apollonius_of_Perga
en.wikipedia.org/wiki/Leonhard_Euler

4.2 Notable Shapes

Roundest Countries

Look around and you will find different circle-shaped objects like our sun, wheels, slices of a watermelon, just to name a few. Is there any country on the map that has such a round shape? Oh, yes! The world's roundest country Sierra Leone is followed by Nauru, Zimbabwe, and the Vatican. Sierra Leone is also the 14th most rectangular country.

The Republic of Sierra Leone is a country in West Africa, on the Atlantic Ocean. It is bordered by Liberia to the southeast and Guinea to the north. Its capital city is Freetown. Sierra Leone is a secular nation that provides freedom of thoughts and religion. Three-quarters of its 8 million citizens are Muslims.

Problem. The area of Sierra Leone is 71,740 km^2. Considering a circular shape of Sierra Leone, what would be the largest distance between its two points?
Round your result to the nearest units.

Answer: 302 km

Squarest Countries

A circle and a square are the most common and perfect figures. These shapes can be found around us. The Vatican is the second squarest country following Egypt and followed by Sint Maarten. It is also the fourth most circular country. The Vatican is the smallest state in the world by area and population. The Vatican is the only independent city-state situated inside city Rome, Italy. The

Vatican city-state got its independence from Italy in 1929. Its name is derived from the name of an Etruscan settlement, *Vatica* or *Vaticum*.

Problem. The area of the Vatican is 0.49 km^2. Assuming a circular shape, what would be the largest distance between its two points and the length of the border of the Vatican? Assuming its square shape, what would be the largest distance between its two points and the length of the border of the Vatican? Compare your results with the actual length of the border between the Vatican City and Italy of 3.2 km.

Round all your results to the nearest hundredths.

Answer: 0.79 km, 2.48 km, 0.99 km, 2.8 km

Shapes of Pyramids

Egypt is famous for its pyramids and mysteries surrounded them. Tourists admire these magnificent structures. Scientists, archeologists, and engineers have been puzzled with their construction and suggested various hypotheses. Sadly, none of them can put an end to their debates. Did the pyramids intend to serve only as a burial place? What was the purpose of chambers, large void, high-ceilinged hallways, and tunnels discovered within the Great Pyramid? What were techniques and technology used to build this architectural masterpiece within a short 20-year period? Who were the laborers? How many of them were employed? Why did they have a 20-tonne invisible door that could be open from the inside and was not visible from the outside? How was it possible to move over 2.3 billion heavy stones to make a pyramid? These and other questions are still waiting for the answer.

Problem. The Great Pyramid in Giza, Egypt, is a right pyramid with a square base as shown in Figure 4.10. The base side (edge) is 230 meters, while the slant height (the height of lateral face) is 187 meters.

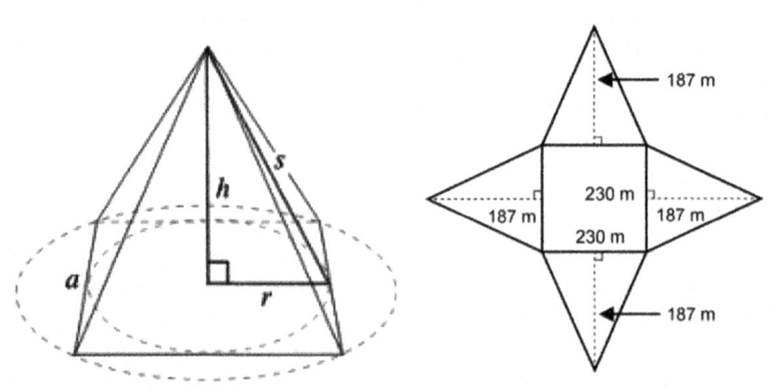

FIGURE 4.10

What are the volume and surface area of the Great Pyramid in Giza? Round your result to the nearest units.

Answer. 138,920 m^2, 2,600,184 m^3

The Pyramid of Giza

The Great Pyramid of Giza is the oldest, most recognizable and the largest pyramid of the three pyramids found in Giza on the west bank of the Nile River. The pyramids are the only item from the list of seven ancient wonders of the world that have survived to our days. In spite of undergoing erosion and major earthquakes during four millennia, the pyramids are still in good condition, though they are not as tall as they used to be when they were first constructed. Shortly after coming to power, Pharaoh Khufu of the eighth dynasty (2589–2566 BCE) ordered it to be built as the grand tomb in which he would be buried. The Great Pyramid of Giza used to be the tallest man-made structure in the world for over 3800 years.

Problem. Find the base edge, height, and slant height of the Great Pyramid in Giza, Egypt, if its surface area is 138,920 m^2 and volume is 2,600,184 m^3.

Note. The Great Pyramid in Giza is wider than taller. Why is this extra information essential? Can the problem be solved without it?

Answer. 230 m, 147.45 m, 187 m, yes, no

Tombs inside Pyramids

Civilizations around the world have built pyramids for religious ceremonies or to bury their monarchs, soldiers, and heroes. The earliest pyramidal structures were built in Mesopotamia, while the Great Pyramid of Cholula in Mexico is the largest pyramid by volume. Sudan has the greatest number of pyramids. The Red Pyramid in the Dashur Necropolis and the Great Pyramid of Khufu in Egypt used to be the world's largest structures for thousands of years and are among the Seven Wonders of the Antiquity.

New Kingdom pharaohs were buried across the Nile in the Valley of the Kings. The finest craftsmen, that lived in the village of Deir el-Medina, built the tombs and decorated the walls and ceilings with magic spells from the Egyptian Book of the Dead, midnight-blue sky, and beautiful scenes. A dead pharaoh was buried with everything he needed for his afterlife as a god.

Problem. A pharaoh's tomb in the Valley of the Kings in Egypt has rectangular shape with a semi round shaped top. It is 2.4 meters tall, 2.6 meters wide, and 5.1 meters long. Its front cross section is depicted in Figure 4.11. Find the volume and surface area of the pharaoh's tomb.

Answer. 28.13 m^3, 56.34 m^2

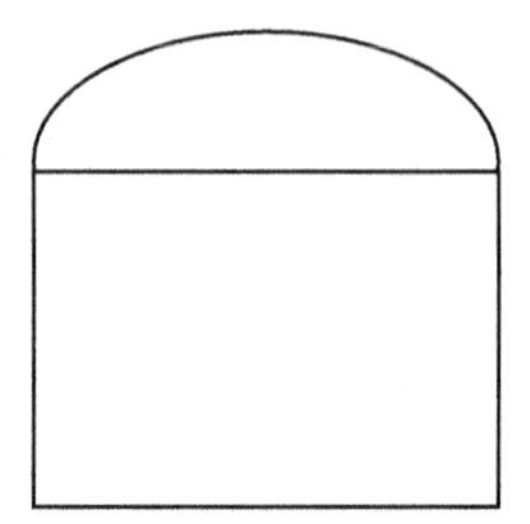

FIGURE 4.11

The Grand Canal

Starting in Beijing and linking the Yellow River in the north and Yangtze River in the south, the Grand Canal or the Jing–Hang Grand Canal for Chinese is a vast waterway system and the longest canal or artificial river in the world with the total length of 1,776 km or 1,104 mi. The oldest parts of the canal date back to the 5th century BCE. Significant restorations have been made since that time. The Grand Canal has reinforced reunion of Chinese north and south playing an important role in ensuring the country's economic prosperity and stability. It is a major transportation and communication means. Its highest point is reached in the mountains of Shandong, at a summit of 42 m or 138 ft. Thoughtfully built pound lock makes it easier for ships to pass these high elevations. The Crand Canal has been listed as a UNESCO World Heritage Site since 2014.

Problem. Find the number of bridges that cross the Grand Canal and the number of locks it has if

- the number of bridges above the Grand Canal is ten times the number of sides in one regular polygon,
- the number of locks is four times the number of sides in the second regular polygon,
- the second regular polygon has twice as many sides as there are in the first polygon.
- the interior angle of the second polygon is twice the interior angle of the first polygon.

Answer. 60, 24

The Corinth Canal

The idea of using a shortcut canal between the Ionian and Aegean seas through the Isthmus of Corinth went back to the 7th century BCE and was implemented in 1893. Sadly, with fast changes in ship size and maritime shipping, the canal soon became obsolete due to its narrowness, shallowness, and maintenance cost. If the canal were built in Antiquity, it would have greatly enhanced trade, transportation, and communications.

Problem. The length of the Corinth Canal in miles is the same as the number of sides in one regular polygon and its deepest point in meters coincides with the number of sides in the second regular polygon that has twice as many sides as there are in the first polygon. The interior angle of the first polygon is twice the exterior angle of the second polygon. Moreover, the time in minutes needed to cross the canal is the same as 5/9 of the measure of the interior angle of the second polygon How long and deep is the Corinth Canal? How much time is needed to cross the canal?

Answer. 4 mi, 8 m, 1 h 15 min

Buildings with Polygons

Since ancient times designers have been creative suggesting their architectural wonders while utilizing triangles, hexagons, and other basic geometric shapes in their fantastic projects that, sometimes, go above any imagination. Triangular shapes provide stability to a structure inspiring designer of the Eiffel Tower in Paris in France. The Centre for Sustainable Energy Technologies in Ningbo in China and housing complex inspired by Mayan architecture in Mexican City of Merida are just other two examples. From molecules to beehives, the hexagonal shape is the most efficient way of covering a maximum surface, because of its use of the least perimeter of separating walls. Using nature as inspiration, architects widely employ hexagonal grids across as the fronts of glass skyscrapers in Singapore, a department store in Mexico City, or the hexagon base of the North Christian Church in Columbus of Indiana in USA

Problem. Find the base area of two buildings, if

A. The base of one building is a regular hexagon, of the second building is an equilateral triangle.
B. The perimeters of an equilateral triangle and the regular hexagon base are the same.
C. The area of an equilateral triangular base is A.

Answer. $3A/2$

The Largest Pomelo

Pomelo is the largest non-hybrid citrus fruit, native to Southeast Asia. Pomelo tastes like a grapefruit, but with an intense tartness and a prevailing sweetness The fruit is huge, 15–25 cm or 6–10 inches in diameter and weighs 1–2 kg or 2–4 lb. Pomelo is endemic to Malaysia.

Problem. The surface area of the heaviest pomelo is 2315.74 cm^2. Assuming a spherical shape, find its volume in m^3.

Answer. 0.01 m^3

The Heaviest Pomelo

Pomelo is highly nutritious, full of fiber, and rich in antioxidants. It is filled with vitamins and minerals and may help you lose weight. The heaviest pomelo mentioned in the Guinness Book of Records weighs over 5 kg or 11 lb. It was grown in Japan.

Problem. The heaviest pomelo is of ellipsoid shape, round in cross-section with the diameter 27.16 cm. Its length is 23.1 cm. Its volume is

A. 8915.6 cm^3
B. 10478.72 cm^3
C. 178431. 2 cm^3
D. 2627.06 cm^3

Answer. A

Solutions

Roundest Countries. The largest distance between any two points on a circle is its diameter. Using the formula for the area of a circle $A = \pi r^2$, we can find its radius and diameter as

$$71740 = \pi r^2 \Rightarrow r \approx 151.11 \Rightarrow d \approx 302.$$

The largest distance between two points in Sierra Leone is 302 km.

Squarest Countries. The largest distance between any two points on a circle is its diameter. Using the formula for the area of a circle $A = \pi r^2$, we can find its radius and diameter as:

$$0.49 = \pi r^2 \Rightarrow r \approx 0.3949 \Rightarrow d \approx 0.7899.$$

The border of a circle is its circumference $C = 2\pi r = 2\pi \cdot 0.3949 = 2.481$.

Using the formula for the area of a square $A = a^2$, we can find its side as $0.49 = a^2 \Rightarrow a = 0.7$. The border is the perimeter $P = 4a = 2.8$. The largest distance between any two points on a square is a diagonal calculated as

$$d = \sqrt{a^2 + a^2} \Rightarrow d \approx 0.9899.$$

The largest distance between two points in the Vatican is 0.79 km and its border is 2.48 km if a circular shape is taken, and 0.99 km and 2.8 km if a square shape is considered. Both estimates of the border are less than the actual border because the Vatican City is neither a perfect square nor a perfect circle.

Shapes of Pyramids. Let H m and a m be the height and the side edge of the pyramid and h m be its slant height, $h = 187$ m, $a = 230$ m. Then, we get

$$H = \sqrt{h^2 - \left(\frac{a}{2}\right)^2}, \text{ or } H = 147.45.$$

Substituting this value to the formulas of the surface area and volume of a pyramid, we obtain

$$S = a^2 + 4 \cdot \frac{ah}{2} \Rightarrow S = 230^2 + 4 \cdot \frac{230 \cdot 187}{2} = 138920,$$

$$V = \frac{1}{3}a^2 H \Rightarrow V = \frac{1}{3} \cdot 230^2 \cdot 147.45 = 20600184$$

The surface area of the Great Pyramid is 138,920 m^2 and its volume is 2,600,184 m^3.

The Pyramid of Giza. Let H m and a m be the height and the side edge of the pyramid and h m be its slant height. From the formula for the volume of a right pyramid, we obtain that

$$h = \sqrt{H^2 + \left(\frac{a}{2}\right)^2} \text{ and } V = \frac{1}{3}a^2 H \Rightarrow a^2 = \frac{3V}{H}.$$

The formula for the pyramid surface area leads to

$$S = a^2 + 2ah \Rightarrow S = \frac{3V}{H} + 2 \cdot \sqrt{\frac{3V}{H}} \cdot \sqrt{H^2 + \frac{3V}{4H}}.$$

After multiplying both sides by H, the last equation gives

$$SH = 3V + \sqrt{3V} \cdot \sqrt{4H^3 + 3V} \Rightarrow SH - 3V = \sqrt{3V} \cdot \sqrt{4H^3 + 3V},$$

which after squaring both sides produces

$$S^2H^2 - 6SHV + 9V^2 = 3V \cdot 4H^3 + 3V \cdot 3V \Rightarrow 12VH^3 - S^2H^2 + 6SHV = 0.$$

Since the pyramid height is nonzero, $H \neq 0$, both sides of the last equation can be divided by H to get the quadratic equation

$$12VH^2 - S^2H + 6SV = 0,$$

which can be solved as

$$H = \frac{S^2 \pm \sqrt{S^4 - 4 \cdot 6SV \cdot 12V}}{2 \cdot 12V} \Rightarrow H = \frac{138920^2 \pm \sqrt{138920^4 - 288 \cdot 138920 \cdot 2600184^2}}{24 \cdot 2600184},$$

or H=147.45 m with the corresponding a = 230 m and H = 471.048 m with the corresponding a = 128.69 m, that is, the quadratic equation produces two different solutions. Because the pyramid is wider than it is tall, the first option should be chosen.

The base side is 230 m, the height is 147.45 m, and the slant height is 187 m.

Tombs inside Pyramids. The pharaoh's tomb consists of two sections: a parallelepiped and a semicylinder. The width of the tomb is the diameter of a semiround shaped top, that is, the radius of the circle is 2.6/2 = 1.3. Then the height of the parallelepiped is 2.4 – 1.3 = 1.1. Its volume is height times width times length, that is, $V = 5.1 \cdot 1.1 \cdot 2.6 \approx 14.59$. The volume of the top semicylinder is $V = \pi r^2 l / 2 = \pi \cdot 1.3^2 \cdot 5.1/2 \approx 13.54$. Finally, the total volume is 14.59 + 13.54 ≈ 28.13.

The surface area consists of sides, bottom, and rounded top and is calculated as

$$S = 2 \cdot 5.1 \cdot 1.1 + 2 \cdot 2.6 \cdot 1.1 + 5.1 \cdot 2.6 + \pi \cdot 1.3^2 + \pi \cdot 1.3 \cdot 5.1 \approx 56.34.$$

The volume and surface area of the pharaoh's tomb in the Valley of the Kings in Egypt are 28.13 m^3 and 56.34 m^2.

The Grand Canal. Let n and N be the numbers of sides in the first and second regular polygons, $N = 2n$. Considering the formula for the interior angle of a regular polygon $\frac{180^0(n-2)}{n}$, we get

$$2 \cdot \frac{180^0(n-2)}{n} = \frac{180^0(2n-2)}{2n}, \text{ from which follows } n = 3, N = 6.$$

Therefore, there are 60 bridges cross the Grand Canal, which has 24 locks.

The Corinth Canal. Let n and N be the numbers of sides in the first and second regular polygons, $N = 2n$. An interior angle of a regular polygon is $\frac{180^0(n-2)}{n}$ and its exterior angle is $\frac{360^0}{n}$.

Since the measure of the interior angle of the first polygon is twice the measure of the exterior angle of the second polygon and $N = 2n$, we can write

$$\frac{180^0(n-2)}{n} = \frac{360^0}{N} \cdot 2 = \frac{360^0}{2n} \cdot 2,$$

from which $n = 4$ and $N = 8$ follows. The interior angle of the second polygon is $\frac{180^0(8-2)}{8} = 135^0$. Then the canal crossing time is $\frac{5}{9} \cdot 135^0 = 75$ minutes or 1 hour and 15 minutes.

The Corinth Canal is 4 miles long and 8 meters deep.

Buildings with Polygons. Let a and b represent sides and A and B stand for the areas of the equilateral triangle and a regular hexagon. Hints A and B lead to $a = 2b$. Connecting middle points of sides of the equilateral triangle divides it into four equilateral triangles, while all line segments that connect vertices of a regular hexagon are equal and intersect at one point forming six equal equilateral triangles, see Figure 4.12. Indeed, the interior angle of a regular hexagon $\frac{180^0(6-2)}{6} = 120^0$divided by 2 is 60^0. All small triangles are equal.

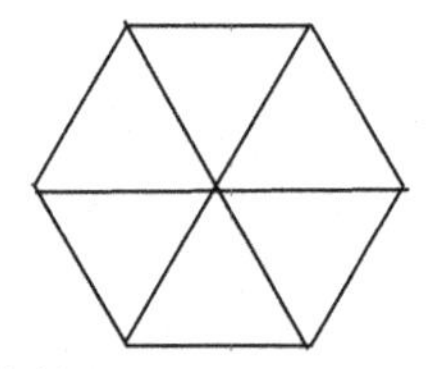

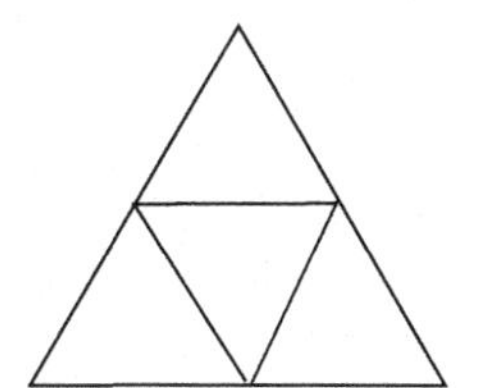

FIGURE 4.12

1st way. The area of an equilateral triangle is $A = a^2\frac{\sqrt{3}}{4}$, and the area of the regular hexagon is

$$B = 6\cdot b^2\frac{\sqrt{3}}{4} = 6\cdot\left(\frac{a}{2}\right)^2\frac{\sqrt{3}}{4} = 6\cdot a^2\frac{\sqrt{3}}{4}\cdot\frac{1}{4} = \frac{3}{2}A.$$

2nd way. Since the area of a triangle is A, then the area of each small triangle is $A/4$. The hexagon has six such triangles. Thus, its area is six times $A/4$ or $3A/2$.

The Largest Pomelo. Using the formulas for the surface area of the sphere $S = 4\pi r^2 = \pi d^2$and its volume $V = \frac{4}{3}\pi r^3 = \frac{1}{6}\pi d^3$, we can get $V = \frac{1}{6}\pi\left(\sqrt{\frac{S}{\pi}}\right)^3$.

Thus, the volume of the pomelo is $V = \frac{1}{6}\pi\left(\sqrt{\frac{2315.74}{\pi}}\right)^3 = 10478.72$.

The volume of the heaviest pomelo is 10478.72 cm^3.

The Heaviest Pomelo. The formulas for the volume for an ellipsoid is $V = \frac{4}{3}\pi abc$, where a, b, and c are its axis, leads to $V = \frac{4}{3}\pi\cdot\frac{27.15}{2}\cdot\frac{27.15}{2}\cdot\frac{23.1}{2}$ $= 8915.6$. Thus, the correct answer is A.

The volume of the heaviest pomelo is 8915.6 cm^3.

Sources

en.wikipedia.org/wiki/Sierra_Leone

citymonitor.ai/community/whats-roundest-country-earth

www.google.com/search?q=Sierra+Leona

www.indy100.com/news/these-are-the-most-rectangular-countries-in-the-world

bigthink.com/strange-maps/sierra-leone-is-the-worlds-roundest-country-and-egypt-the-squarest-one/

en.wikipedia.org/wiki/Vatican_City

www.worldatlas.com/articles/10-astounding-facts-about-the-great-pyramid-of-giza.html

en.wikipedia.org/wiki/Pyramid

whc.unesco.org/en/list/1443

en.wikipedia.org/wiki/Grand_Canal_(China)

www.worldatlas.com/articles/10-astounding-facts-about-the-great-pyramid-of-giza.html

www.dezeen.com/2018/07/26/six-buildings-hexagon-pattern-honeycomb-facades/

en.wikipedia.org/wiki/Pomelo facts.net/country-facts/

www.guinnessworldrecords.com/world-records/heaviest-pummelopomelo

en.wikipedia.org/wiki/Pomelo

4.3 Geometric Fun with National Flags

Colorful National Flags

The national flag is one of the most important and recognizable symbols of a country. The flag design and colors feature a country's history, heritage, culture, religion, and other distinctive characteristics. The color white often symbolizes social justice and peace, green is for the land, agriculture, and belonging to Islam, the color red is used to signify the fight for independence and the courage of citizens, blue is for the sky and the sea, yellow is for the sun, and so on. The flag of the only English-speaking Central American country, Belize, is considered to be the world's most colorful flag, though the majority of its 12 colors are for the coat of arms. If only colors on a primary design are considered, then the national flag of South Africa is the most colorful flag with six core colors. The Central African Republic, Comoros, and Seychelles use five primary colors in their flags shown in Figure 4.13.

Problem. Blue, yellow, red, white, and green rays alternate on the national flag of Seychelles. The rays go from the lower left corner up dividing the opposite side into three equal parts. Blue, yellow, white, and green rays form triangles. The middle red ray is a quadrilateral for it ends at the top right corner taking the third of the width and the third of the length. The proportion of the rectangular flag is 1 to 2. Express the area of each colorful section in terms of the width w of a flag.

Answer. $w^2/3$, $2w^2/3$

FIGURE 4.13

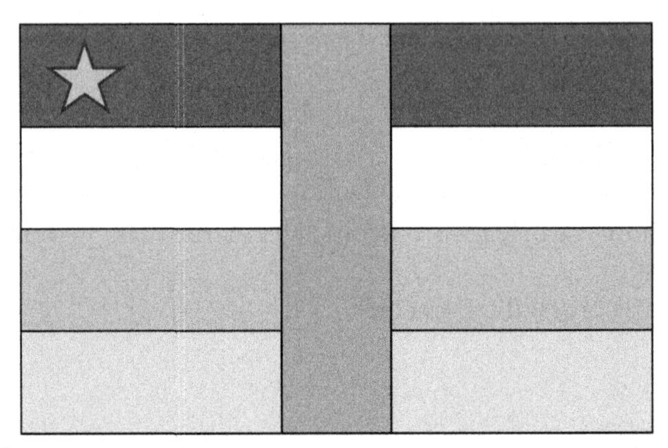

FIGURE 4.14

The Unity Flag

The Central African Republic is a Central African landlocked country. It borders another landlocked country Chad, the world's newest country South Sudan, Sudan with more pyramids than Egypt, French-speaking multiverse DR Congo, Republic of Congo, and the wettest country Cameroon. Rich in diamonds, gold, oil, timber, and other natural resources, the Central African Republic is one of the poorest countries in the world. This multi-cultural country has used its five-color flag since 1958. Its four horizontal stripes echo the blue and white colors of France and green and yellow colors of Africa. The red vertical stripe highlights the unity that Europeans and Africans should have for each other.

Problem. The Constitution of the Central African Republic states that the national flag is of a rectangular shape with four equal blue, white, green, and yellow horizontal bands, perpendicularly barred in their center by a red band. The proportion of the flag is 2 to 3. The smallest sides of the vertical and horizontal stripes are the same. Find the ratio of areas of white and red bands.

Answer. 5:4

Rectangles on National Flags

The West African powerful Dahomey Kingdom gave birth to Benin that took its name from the Bay of Benin. With 17 years as a median age of its population, this French-speaking country has the youngest population in the world. It is also a place where the vodun or voodoo religion originated. It is believed there that you will be lucky if a snake crosses your path.

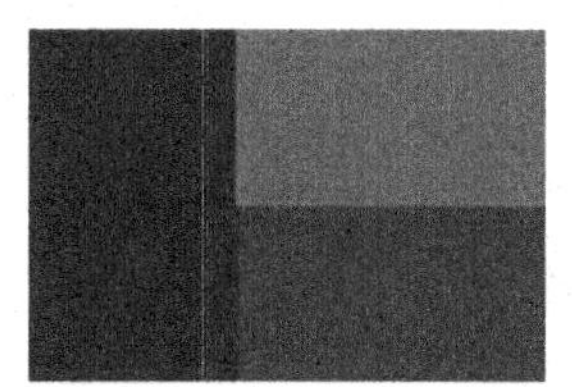

FIGURE 4.15

The national flag of Dahomey adopted in 1959 to replace the French Tricolour, was changed in 1975, and then reinstated. Like Ethiopia, Ghana, Cameroon, Guinea, and many other neighboring countries, the national flag has red–yellow–green pan-African colors. Yellow and green stand for savannas and palm groves, while red represents the courage of the ancestors who defended the homeland.

Problem. The nation flag of Benin (see Figure 4.15) consists of a vertical green band on a hoist and two equal yellow and red horizontal bands on the fly side. While sewing the national flag, students considered the following measurements.

A. The ratio of the width (fly) of the nation flag of Benin to its length (fringe) is 3 to 4.
B. The flag is 5 cm longer than wider.
C. The ratio of areas of its yellow and green bands is 3 to 4.

Find the size of each color band on the nation flag of Benin.

Answer. 6 by 10, 9 by 5, 9 by 5

Triangles on National Flags

Triangles are often used in flag design. Three triangles set up the background of the national flag of Eritrea. The centered isosceles red triangle is based on the host side with the vertex on the opposite side, while right green and blue triangles are faced in the opposite direction. Though having different colors and emblems, the three triangles of the national flag of American Samoa can be seen as the mirror reflection of the Eritrean flag. Two right triangles, black and red, with Raggiana bird-of-paradise and the Southern Cross attract attention to the flag of Papua New Guinea. The flag of Guyana, called *the Golden Arrowhead,* features six triangles.

Problem. Find the ratio between the areas of two isosceles triangles. One triangle is based on the host side with the vertex on the opposite side. The

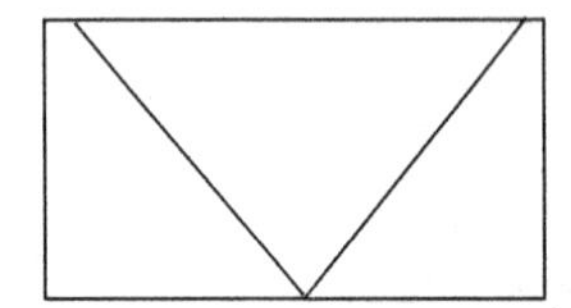

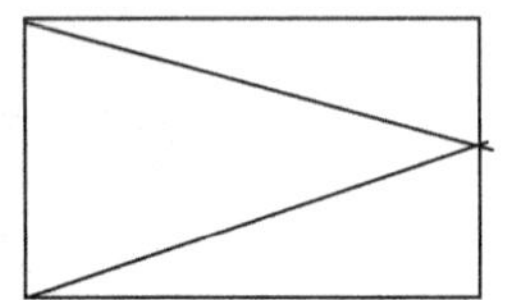

FIGURE 4.16

second triangle is based on the fly side with the vertex on the opposite (fringe) side. Both flags have the same dimensions. What would be the ratio if one of the triangles is a right or scalene triangle?

Answer. 1, the same

National Flags with Triangles

Triangles of different colors and shapes are commonly used in a flag design. A horizontal tri-color isosceles triangle on the national flag of Antigua and Barbuda and a yellow right triangle of Bosnia and Herzegovinian flag are based on the top edge, while two triangles are centered on the Saint Lucian flag. Triangles are based at the hoist of most national flags. The national flags of Philippines and Djibouti have a white equilateral triangle, a red one is on the Cuban flag, and a blue triangle is on the flags of São Tomé and Príncipe and Equatorial Guinea, while the national flag of the Czech Republic has a blue isosceles triangle.

Problem. The height of an isosceles triangle on the flag of São Tomé and Príncipe is a half of the width (hoist) of the flag, while the height of an isosceles triangle on the flag of Jordan is the same as the width of the flag. A black equilateral triangle is on the flag of East Timor. The national flag of which country has a triangle with the largest area? What should be the ratio between measurements in an isosceles triangle to have its area greater than the area of an equilateral triangle with the same base? The length of all hoists is the same.

Answer. $h > \frac{w\sqrt{3}}{2}$, no

Trapezoids and Triangles on National Flags

A flag with a black triangle situated at the hoist and a horizontal gold band surrounded with two aquamarine horizontal bands has been the national flag of the Commonwealth of the Bahamas since the country gained its independence on 10 July 1973 after being a crown colony of the United Kingdom

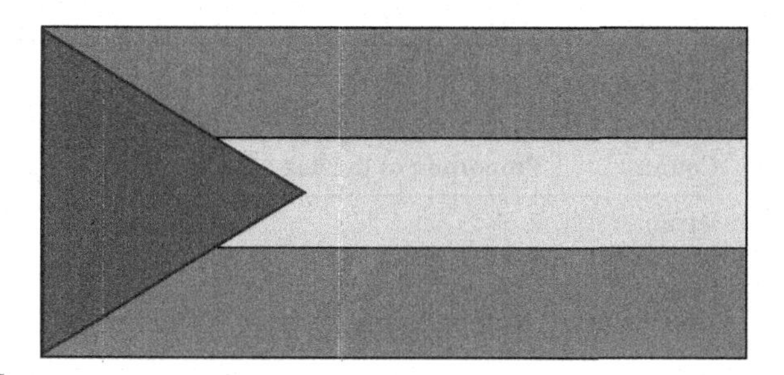

FIGURE 4.17

for over 250 years. It also changed its name from the Bahama Islands to the Bahamas. The gold on the Bahamas flag indicates its natural resources and the shining sun. Aquamarine color emphasizes cleanest water that surrounds 700 Bahamas islands. The black signifies the strength of the Bahamians.

Problem. The national flag of the Bahamas consists of a black equilateral triangle and the gold stripe between two aquamarine horizontal stripes of the same width. The proportion of the flag width to its length is 1:2. Present the areas of a triangle and each stripe in terms of the width.

Answer. $\frac{\sqrt{3}}{4}w^2, \frac{24-5\sqrt{3}}{36}w^2, \frac{12-\sqrt{3}}{18}w^2$

Circles on National Flags

Despite different colors, national flags of Japan, Bangladesh, and Palau have similar designs with a circle at the center on a solid-color background.

The Japanese flag *Hinomaru* or officially *Nisshōki*, meaning the *sun-mark flag*, has a red circle on a pure white background that reflects purity and honesty. Legends say that the first emperors of the Land of the Rising Sun were descendants of the Sun Goddess.

The red circle on the flag of Bangladesh represents both the sun rising over a new nation and the blood spilled during the struggle for independence. A dark green background portrays the beauty of the land and devotion to Islam.

The golden disk on the Palauan national flag represents the yellow moon that highlights peace and achievements of the nation and provides the feeling of warmth and love. Surprisingly, its blue background is not related to the Pacific Ocean, but rather honors the independence of the Palau Island.

Problem. A shaded circle is central on the national flags of Japan, Bangladesh, and Palau. The proportions of flags and dimensions of the circles are presented in the table below. What circle has the largest area if the hoists of all flags are the same (all flags have the same width)?

Country	Proportion of the flag	Circle
Japan	2: 3	$d = 6/5$
Bangladesh	1: 5/3	$d = 2/3$
Palau	5: 8	$d = 3$

Answer. The flag of Bangladesh.

National Flags with Circles

The national flag, seal, and many other official symbols represent a country and are specific to each country, signify its essence, and bring a specific message. The national flags of Japan, Bangladesh, and Palau have similar and relatively simple designs featuring a circle on a solid color background, though they were adopted in different years, in 1810, 1872, and 1981 correspondingly.

Problem. Despite different colors and width / length ratio of flags, a shaded circle on the national flags of Japan, Bangladesh, and Palau is their distinctive feature. The flags and their dimensions are presented in Figure 4.18. What nation designates the largest area for a circle on its nation flag?

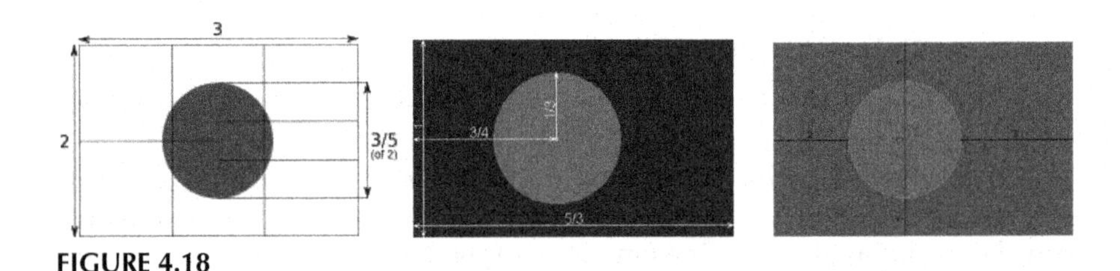

FIGURE 4.18

Answer. The flag of Bangladesh.

Five-Pointed Stars on National Flags

National flags of Philippines, Venezuela, Cuba, Panama, Jordan, Pakistan, and many other countries have five-pointed stars of various colors and quantities put on different backgrounds. Ghana and São Tomé and Príncipe have black stars on their flags, while stars on the Syrian and Iraqi flags are green and on the flags of Algeria and Djibouti are red. Australia and Kosovo have white stars on their flags, while Panama has red and blue stars. Fifty white five-pointed stars on the current United States' flag represent the unity of its 50 states, the five blue stars on the flag of Honduras stand for five nations of the United Provinces of Central America, El Salvador, Costa Rica, Nicaragua, Honduras, and Guatemala.

Problem. The diameter of a circumscribed circle of a five-pointed star on the national flag is d cm. What is the area of one star? Round to the nearest hundredth.

Answer. $0.28d^2$ cm^2

Six-Pointed Stars on National Flags

The Republic of Burundi in East Africa is bordered by Rwanda, Tanzania, and the Democratic Republic of the Congo. Through its rich history, Burundi has been an independent kingdom, a part of German East Africa and of Ruanda-Urundi, the territory to Belgium, a monarchy, and, finally, a republic.

The current Burundian flag was adopted on March 28, 1967, after the country gained its independence from Belgium. Burundi has modified its design more often than any other African nation. A white saltire with three stars divides the flag into alternating green and red sections to symbolize hope for the future and blood of those who struggled for independence. The white stands for peace and harmony. The red six-pointed stars represent both the three major ethnic groups of Burundi, the Hutu, Twa, and Tutsi, and three elements of the national motto, *Unity, Work, and Progress.*

Problem. The Burundi flag centers a white circle with three six-pointed identical stars (see Figure 4.19). What is the area of a star if the diameter of a circumscribed circle of the star is 20 cm?

Answer. $100\sqrt{3}$

FIGURE 4.19

Solutions

Colorful National Flags. Let l denote the length and w stand for the width of a flag, $l = 2w$. Let us draw a diagonal from the left low corner to the right upper corner of the flag. Blue, yellow, and the upper red triangles have equal bases, that is, $l/3 = 2w/3$, and the same height w. Thus, their areas are the same and equal to

$$\frac{w \cdot l/3}{2} = \frac{w \cdot 2w/3}{2} = \frac{w^2}{3}.$$

Green, white, and the lower red triangles have equal bases $w/3$, and the same height $l = 2w$. Therefore, their areas are the same and equal to

$$\frac{l \cdot w/3}{2} = \frac{2w \cdot w/3}{2} = \frac{w^2}{3}.$$

The red part consists of two triangles, that is, its area is twice $w^2/3$ or $2w^2/3$.

Blue, yellow, white, and green rays on the national flag of Seychelles have the same areas of $w^2/3$ square units. The area of a red ray is $2w^2/3$ square units.

The Unity Flag. Let l denote the length and w stand for the width of a flag, $l = \frac{3}{2}w$. The smallest side of the red stripe is $w/4$, while its longest side is w. The blue, white, green, and yellow horizontal bands are the same, that is, their width is $w/4$, while its longest side is $l - w/4 = 3w/2 - w/4 = 5w/4$. Thus, the area of the white horizontal stripe is $\frac{5w}{4} \cdot \frac{w}{4} = \frac{5w^2}{16}$, and of the red vertical stripe is $\frac{w}{4} \cdot w = \frac{w^2}{4}$. Hence their ratio is expressed as $\frac{5w^2}{16} \div \frac{w^2}{4} = \frac{5}{4}$.

The ratio of areas of white and red stripes is 5 to 4.

Rectangles on National Flags. Let x be the length of the flag, then, from Hint B, its width is $x - 5$. Following Hint A, we get

$$\frac{x-5}{x} = \frac{2}{3} \Rightarrow 3(x-5) = 2x \Rightarrow x = 15.$$

Let y be the length of a yellow section of the flag, then its area is $\frac{10}{2} \cdot y$, and the area of the green segment is $(15 - y) \cdot 10$. Using Hint C, we obtain

$$\frac{5y}{10 \cdot (15-y)} = \frac{3}{4} \Rightarrow y = 9.$$

The red and yellow bands of the national flag of Benin are 9 cm by 5 cm, its green band is 6 cm by 10 cm.

Triangles on National Flags. The area of a triangle is the height times the base over two. If a triangle is based on the host side with the vertex on the opposite side, then the width of a flag is the base of a triangle, and its height is the length of the flag. If a triangle is based on the fly side with the vertex on the opposite side, then the width of a flag is the height of a triangle, and its length is the base of the flag. Thus, the same numbers, the width and the length, are multiplied and divided by two to find the areas of the triangles. Hence, the areas are the same. The result does not alternate if right or scalene triangles instead of isosceles triangles are taken. Therefore, the ratio of any pair of these areas is 1.

National Flags with Triangles. The area of an equilateral triangle is $A_{eq} = \frac{w^2\sqrt{3}}{4}$ and the area of an isosceles triangle is $A_{is} = \frac{wh}{2}$. The triangles are based on the hoist-side with the same flag width w. The area of an isosceles triangle is greater than the area of an equilateral triangle if

$$A_{eq} < A_{is} \text{ if } \frac{w^2\sqrt{3}}{4} < \frac{wh}{2} \text{ or } h > \frac{w\sqrt{3}}{2}.$$

For the flag of São Tomé and Príncipe$h = \frac{w}{2} < \frac{w\sqrt{3}}{2}$, while for the flag of Jordan $h = w > \frac{w\sqrt{3}}{2}$.

The area of a triangle on the flag of Jordan is the largest among the three.

Trapezoids and Triangles on National Flags. Let w and $l = 2w$ be the width and length of the flag. The area of one third of a rectangular flag is $\frac{1}{3}lw = \frac{2}{3}w^2$. The area of ΔABC is $\frac{\sqrt{3}}{4}w^2$ and of ΔABR is $\frac{1}{2} \cdot \frac{\sqrt{3}}{4}w^2 = \frac{\sqrt{3}}{8}w^2$ (see Figure 4.20).

Triangles ΔABR and ΔAQP are similar, $AR = \frac{1}{2}w$, $AP = \frac{1}{3}w$, $BR = \frac{\sqrt{3}}{2}w$, then $PQ = \frac{\sqrt{3}}{3}w$ and the area of a ΔAPQ is $\frac{1}{2} \cdot \frac{\sqrt{3}}{3}w \cdot \frac{1}{3}w = \frac{\sqrt{3}}{18}w^2$, which subtracted from the area of one third of a rectangular flag gives the area of one aquamarine stripe $A_{aq} = \frac{2}{3}w^2 - \frac{\sqrt{3}}{18}w^2 = \frac{12-\sqrt{3}}{18}w^2$.

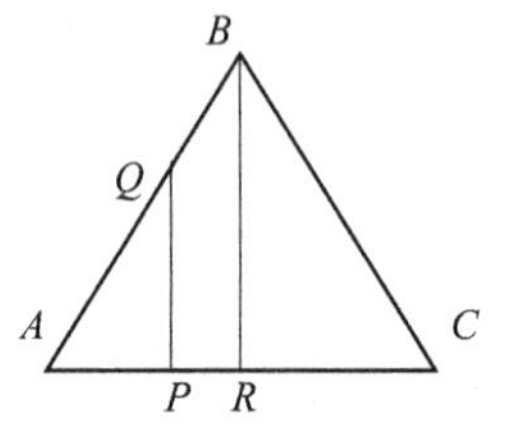

FIGURE 4.20

The area of the trapezoid $QPRB$ is $A_{\Delta ABR} - A_{\Delta APQ} = \frac{\sqrt{3}}{8}w^2 - \frac{\sqrt{3}}{18}w^2 = \frac{5\sqrt{3}}{72}w^2$, which after multiplication by 2 and subtraction from the area of one third of a rectangular flag gives the area of one gold stripe $A_{gold} = \frac{2}{3}w^2 - \frac{5\sqrt{3}}{36}w^2 = \frac{24-5\sqrt{3}}{36}w^2$.

Circles on National Flags. A larger circle has a larger diameter. Since the provided flag dimensions are given in different counts, the diameters of the circles should be compared on the flags with the same hoist (width).

The diameter of the red circle is 6/5 units on the Japanese flag if the flag hoist (width) is 2 units. The diameter of the red circle is 2/3 units on the Bangladeshis flag if the flag width is 1 unit, or 4/3 if its width is 2 units. The diameter of the yellow circle is 3 units on the Palauan flag with the flag width is 5 units, or 6/5 if the width is adjusted to 2 units. Hence, we get:

$\frac{4}{3} > \frac{6}{5}$ that follows from writing the fractions with a common denominator, $\frac{20}{15} > \frac{18}{15}$.

The circle on the Bangladesh flag has the largest diameter, while diameters of the circles on the Japanese and Palauan flags are the same.

National Flags with Circles. Considering the area of a circle $A = \pi r^2$ and of a rectangle $A = width \cdot length$, the goal is to compare ratios between these two areas for each flag, that is

$\frac{\pi \cdot \left(\frac{3}{2\cdot 5}\right)^2}{2\cdot 3} = \frac{\pi \cdot 3}{200}$ for the flag of Japan, $\frac{\pi \cdot \left(\frac{1}{3}\right)^2}{1\cdot \frac{5}{3}} = \frac{\pi}{15}$ for the flag of Bangladesh, and

$$\frac{\pi\cdot\left(\frac{3}{2}\right)^2}{5\cdot 8}=\frac{\pi\cdot 9}{160} \text{ for the flag of Palau.}$$

Writing the fractions with a common denominator and comparing them, we get:

$$\frac{\pi}{15}=\frac{160\pi}{2400}>\frac{9\pi}{160}=\frac{135\pi}{2400}>\frac{3\pi}{200}=\frac{12\pi}{2400}.$$

The circle on the Bangladeshis flag takes up the largest area of the flag.

Five-Pointed Stars on National Flags. Let r be the radius of a circle, so $r = d/2$. Then,

$$\angle OAC = 360° / 10 = 36°, \text{and } \angle OAC = (360^0 / 10) / 2 = 18^0$$

in the triangle ΔAOCin Figure 4.21:

1st way: by the Theorem of Sines, $\frac{r}{\sin 126^0}=\frac{OC}{\sin 18^0}=\frac{AC}{\sin 36^0}$. By the Heron's formula, $A_{\Delta AOC}=\sqrt{p(p-r)(p-\frac{r\sin 18^0}{\sin 126^0})(p-\frac{r\sin 36^0}{\sin 126^0})}=0.112257r^2$, where p is a semi perimeter of ΔAOC, $p=\frac{1}{2}\left(r+\frac{r\sin 18^0}{\sin 126^0}+\frac{r\sin 36^0}{\sin 126^0}\right)$.

Then $A_{star}=10\sqrt{p(p-r)(p-\frac{r\sin 18^0}{\sin 126^0})(p-\frac{r\sin 36^0}{\sin 126^0})}= 1.12257\, r^2 \approx 0.28d^2$.

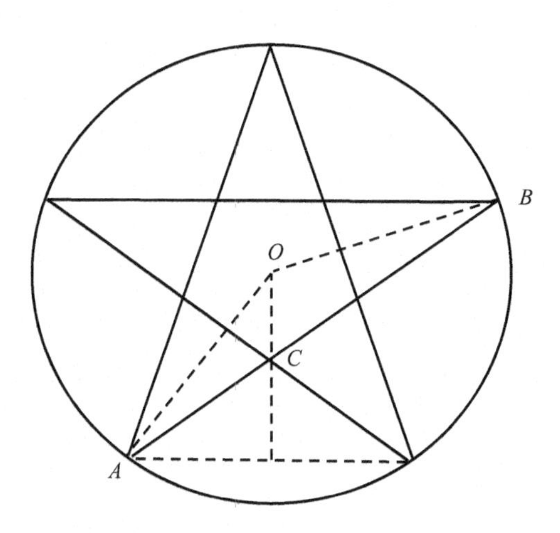

FIGURE 4.21

2nd way: by the Theorem of Sines, $\frac{r}{\sin 126^0} = \frac{OC}{\sin 18^0}$, therefore, the height to AO in ΔAOC

$$h = OC \sin \angle AOC = \frac{r}{\sin 126^0} \sin 18^0 \cdot \sin 36^0.$$

Then, $A_{\Delta AOC} = \frac{rh}{2} = \frac{r^2}{2\sin 126^0} \sin 18^0 \cdot \sin 36^0$ and $A_{star} = \frac{5r^2}{\sin 126^0} \sin 18^0 \cdot \sin 36^0$ $= 1.12257\, r^2 \approx 0.28d^2$.

3rd way: The area of ΔAOB is $A_{\Delta AOB} = r^2 \sin 18^o \cos 18^o$. Five triangles like ΔAOB form the star, but the inner pentagon is taken twice, its area $A_{pentagon} = 5r^2 \sin^2 18^o \tan 36^o$. Then

$$A_{star} = 5A_{\Delta AOC} - A_{pentagen} = 5 \cdot r^2 \sin 18° \cos 18° - 5r^2 \sin^2 18° \tan 36°$$
$$= 1.12257r^2 \approx 0.28d^2.$$

There are other ways to solve this problem.

The area of a pentagram on the flag is $0.28d^2$ cm^2.

Six-Pointed Stars on National Flags. The circum radius $r = 20/2 = 10$ cm. In the triangle ΔAOC (see Figure 4.22), $\angle AOC = 360^0 / 12 = 30^0$; $\angle OAC = \frac{1}{2}\angle POQ =$ $(360^0 / 6) / 2 = 30^0$.

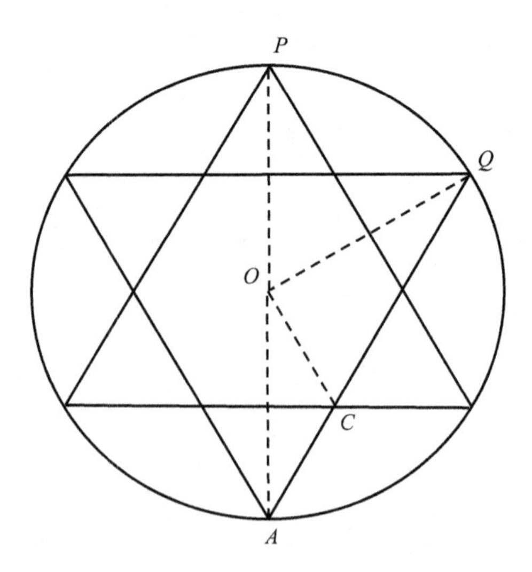

FIGURE 4.22

Therefore, $OC=CA$, the height to OA in ΔAOC is $h = \frac{r}{2} \cdot \tan \angle AOC = \frac{r\sqrt{3}}{6}$, and the area of the triangle ΔAOC is $A = \frac{rh}{2} = \frac{r^2\sqrt{3}}{12} = \frac{25\sqrt{3}}{4}$. Hence, $A_{star} = 12 \cdot \frac{r^2\sqrt{3}}{12}$ $= 100\sqrt{3}\,\text{cm}^2$.

Sources

flagpoles.co.uk/flag-geek-which-national-flag-has-the-most-colours

www.allstarflags.com/facts/worlds-ten-most-colorful-flags/

en.wikipedia.org/wiki/Flag_of_the_Central_African_Republic

www.enjoytravel.com/en/travel-news/interesting-facts/interesting-facts-about-central-african-republic

www.britannica.com/topic/flag-of-Benin

en.wikipedia.org/wiki/Flag_of_Benin

www.enjoytravel.com/us/travel-news/interesting-facts/interesting-facts-benin

en.wikipedia.org/wiki/List_of_national_flags_by_design

en.wikipedia.org/wiki/Flag_of_Guyana

en.wikipedia.org/wiki/Flag_of_American_Samoa

en.wikipedia.org/wiki/Flag_of_Papua_New_Guinea

en.wikipedia.org/wiki/Flag_of_the_Bahamas

www.lonelyplanet.com/articles/flag-of-japan-facts-history-and-design

www.britannica.com/topic/flag-of-Palau

www.lonelyplanet.com/articles/flag-of-palau-facts-history-and-design

www.lonelyplanet.com/articles/flag-of-bangladesh-facts-history-and-design

www.worldometers.info/geography/flags-of-the-world/

wikimedia.org/wiki/Flags_with_stars

www.gettysburgflag.com/flags-banners/burundi-flags,

en.wikipedia.org/wiki/Burundi

5

The World in Puzzles

> The most enjoyable part is the puzzle, the process of solving, not the solution itself
>
> ~ Erno Rubik (1944–)[1]

We exercise our bodies, but what's about challenging our brains and giving them a workout to sharpen memory, boost creativity, and improve concentration? What can be better than working on puzzles that will also take your stress and anxiety away?

Doing sudoku, kakuro, inky, magic square, logic, pattern, de-coding, and 40 similar amazing puzzles offered in this chapter will put you in a pleasant, meditative state. These puzzles are designed to stimulate diverse intellectual and logical skills in addition to reviewing mathematical concepts. Some of these puzzles will be quick and easy, while others will test your patience and wits, but all of them will reward you with a relaxed and triumphant feeling after their successful completion! At the same time, you will learn surprising facts about geography, heritage, and history around the globe and inventions made by geniuses. Grab a warm beverage, find a comfy seat, and enjoy!

5.1 Savvy Puzzles

Countries Named Guinea

Guinea is a region in West Africa named after the Gulf of Guinea. French Guinea, Spanish Guinea, Portuguese Guinea, German Guinea, and other former European colonies there were named after the gulf. There are many theories on where the word *Guinea* came from. Probably, it is derived from the Portuguese word *Guiné* originated during the 15th century to refer to the lands owned by African people Guineus who came from regions south of the Senegal River. Another theory claims that *Guinea* is derived from a city in modern-day Mali, Djenné, important during the trans-Saharan trade in the

[1] A Hungarian designer and architect, who invented the Rubik's Cube and Rubik's Magic.

 DOI: 10.1201/9781003351702-5

15th to 17th centuries. Or, perhaps, the word came from the Arabized word for *Ghinawen*, meaning *blacks*.

Problem. Find how many countries have *Guinea* in their name if this number is a missing number **X** in the following infinite sequence

$$3, 3, 5, 4, \mathbf{X}, 3, 5, 5, \mathbf{X}, 3, 6, \ldots .$$

Name these countries and their location.

Answer. 4

Equatorial Guinea

Equatorial Guinea borders Cameroon and Gabon in West Africa. The country consists of Río Muni in continental Africa and several volcanic islands. Its capital city of Malabo is on one of the islands, though the new capital, Cuidad de la Paz, is currently under construction on the mainland. Surprisingly, despite its gorgeous unique nature, Equatorial Guinea is one of 27 countries that doesn't have any UNESCO World Heritage Sites.

Portugal was the first to colonize islands and later exchanged them with Spain for regions in Latin America. Spanish Guinea gained independence in 1968 giving birth to the Republic of Equatorial Guinea. It is the only Spanish-speaking country in Africa and declares Spanish as its official language.

Problem. How many volcanic islands does Equatorial Guinea have if this number is the only number that appears on three positions determined by three consecutive odd numbers of the finite sequence with 27 one-digit terms

$$1, 0, 9, 2, 1, 8, 3, 2, 7, 4, 3, 6, 5, \ldots, 9, 8, 1?$$

Answer. 5

The Largest Gas Station

Texas is a famous US state at the border with Mexico. The popular saying *Everything is bigger in Texas* reflects the enormous area of this largest state among the contiguous 48 states of the United States. You can drive for almost a day and still be in Texas! Texas maintains the biggest, the largest, the longest, the widest, and other world's records.

Problem. The world's largest Gas Station is Buc-ee's, in New Braunfels in Texas. Find a three-digit number of gas pumps there, if it can be presented as the lowest product of five consequent positive numbers.

Answer. 120

Mont-Saint-Michel

A magnificent island topped by a gravity-defying abbey built atop a mountain is the most unforgettable and breathtaking view. This 4 km^2 rocky island is surrounded by water at high tide. A legend states that Aubert, bishop of town of Avranches in France claimed that Archangel Michael had instructed him to build a church on the top of the *Mont Tombe*. Building the abbey on a sharp summit that raised 92 m above sea level and maintaining it have become an architectural and engineering wonder. Throughout centuries Mont-Saint-Michel has been a pilgrimage site, a center of learning, a refuge, a monastery, a military defense fort, and, even, a prison. Mont-Saint-Michel appears in the UNESCO list of World Heritage Sites.

Problem. When did Mont-Saint-Michel become a World Heritage Site by UNESCO? When did Bishop of Avranches have his dream of building a church on the top of the Mont Tombe? Use the following hints to answer these questions.

A. The difference between ones and tens digits of the year when Mont-Saint-Michel was listed as a UNESCO Heritage Site (UNESCO year for short) is the smallest prime.
B. Two-digit numbers formed from the last digits of the UNESCO year and its first two digits in the order they appeared in the year are primes.
C. The difference between the two-digit numbers described in Hint B can be presented by another two-digit number obtained as the product of three consecutive numbers that are co-primes.
D. The year when Aubert Bishop of Avranches had a dream to build a church on Mont-Tombe is a three-digit number with the hundreds equal to the tens digit of the UNESCO year, the tens presented as a difference between hundreds and units, and its last digit given as a difference between the hundreds and thousands units of the UNESCO year.

Answer. 1979, 708

Coral Reefs

Coral reefs are the most diverse underwater ecosystem that generously supports marine life. These massive, gorgeous sea rainforests are made of limestone deposited by coral polyps. Coral reefs are classified not only by their breathtaking beauty and uniqueness but also by their origin, forms, location, as the fringing or shore reef, barrier reef, atoll, table or platform reef, patch reef, apron reef, and other types. Coral reefs comprise less than 0.1% of the ocean with 91.9% of this number belonging to the Indo-Pacific water region. Located in the Coral Sea, the largest and longest Great Barrier Reef is a system of over 3000 reefs. It is one of Australia's most remarkable natural wonders.

Problem. Find the length of the Great Coral Barrier Reef of Australia in miles if

A. Its length is a four-digit number.
B. The first, second, and fourth digits of the length are squares of three numbers in increasing order.
C. The third (tens) digit is the difference between the square roots of the fourth and first digits.
D. All digits are different.

Answer. 1429 mi

Three Geniuses

Mohandas Karamchand Gandhi was born in India, educated in Great Britain, and worked in South Africa for two decades before coming back to his native country, where he led anti-colonial nonviolent resistance to British rule in India. He was nominated five times for the prestigious Nobel Peace Prize and would have received it in 1948 if he had not been assassinated. *Time* magazine named Mahatma Gandhi *Person of the Year* in 1930. The United Nations declared Gandhi's birthday, October 2, as the International Day of Non-Violence in 2007.

Henri Émile Benoît Matisse gave up his career as a lawyer to study art. While developing his own style, he became one of the most recognizable French visual artists known for using bright colors to express his ideas and making revolutionary development in the arts.

Valdemar Poulsen is a Danish engineer that, while working for the Copenhagen Telephone company, was the first one to demonstrate that sound could be recorded magnetically. His invention was the forerunner of the tape recorder. He also worked on the development of long-wave radio broadcasting.

What do an Indian leader, a Danish engineer, and a French impressionist have in common? Surprisingly, they were born in the same year.

Problem. Using the hints below, find the year when Mahatma Gandhi, Henri Matisse, and Valdemar Poulsen were born.

A. The first (thousands) and the last digits are squares.
B. The first and the second digits are cubes.
C. The third digit is the product of the fourth root of the first digit, the cube root of the second digit, and the square root of the fourth digit.
D. None of the digits are repeated.

Answer. 1869

The Braille System

Vision is one of the most essential human sensors. Blindness... People cannot see but should they give up? Turkish artist Eserf Armagan, American YouTuber and film critic Tommy Edison known as the *Blind Film Critic,* Japanese tenor Tsutomu, and American artist Jeff Hanson were born blind or became blind at a very young age, but blindness did not prevent them to teach themselves and succeed in what they chose. American athlete Erik Weihenmayer was the first blind person to reach the summit of Mount Everest. Blind Ancient Greek orator Homer authored *Homer* and *Odyssey*. Famous painters French impressionist Claude Monet and Dutch artist Rembrandt Van Rijn became blind later in their lives but continued to work. Blindness since the age of 59 did not prevent Swiss mathematician and physicist Leonhard Euler from making complicated calculations in his head and dictating his books in mathematics.

Lous Braille was born in Paris, France. He became blind at the age of three. Teaching at a school for the blind, he developed a system of six dots arranged in two parallel rows that could be raised to produce 64 combinations to represent letters and words that can be read with the fingers. As a talented musician, he also developed a Braille musical codification. The Braille system adjusted to other native languages has been used by visually impaired people around the world.

Problem. Using the following hints, find when Lous Braille was born. The first two digits of the year of his birth form the first two-digit number, the last two digits of the year form the second two-digit number.

A. The sums of digits in both numbers are the same.
B. The first number is twice the second number.

C. If digits in the first number are reversed, then the obtained number will be the second number squared.

Answer. 1809

NATO

The North Atlantic Treaty Organization (NATO) is an intergovernmental military alliance between European and North American countries which was signed in Washington, DC, on April 4. New countries have joined NATO since that time. NATO countries have agreed to defend each other against attacks by third parties. NATO's main and military headquarters are in Belgium.

Problem. The square root of the second digit of the year when NATO was founded is the difference between its third (tens) and first digits. The square root of the fourth digit of the year is the sum of the square roots of its third and first digits. The product of all digits of the year is nonzero.

The number of founding countries is 1 less than the sum of the tens and units digits of the year, while the number of NATO countries as of 2022 is 3 times the sum of the thousands and units digits of the year, and the number of non-European countries is the square root of the tens digits of the year.

Find the year when NATO was formed, the number of founding countries, the number of countries NATO had in 2022, and the number of NATO countries outside Europe.

Answer. 1949, 12, 30, 2

The World's Fair

St. Louis in Missouri hosted the World's Fair to celebrate the centennial of the Louisiana Purchase, Thomas Jefferson's great accomplishment. Therefore, the fair was called the *Louisiana Purchase Exposition*. It revealed novel technological ideas, consumer goods, and scientific innovations to millions of visitors from over 60 countries and 43 of the then 45 American states.

The exhibitions presented and predicted new technologies that had been developed by that time. Many of them have been modified and used today brightening the everyday life of people. The *radiophone*, invented by Alexander Graham Bell, has been developed into the radio and telephone. The *telautograph*, suggested by Elisha Gray, was the predecessor to the modern fax machine. *Finsen light*, for which Niels Ryberg Finsen was awarded the Nobel Prize, has led to radiation therapy treatment. The *X-ray* machine, proposed earlier by Wilhelm Conrad Röntgen, had its public debut. The *infant incubator*, suggested by Drs. Alan M. Thomas and William Champion, has saved many lives. Thomas Edison had his own exhibition. The electrical plug, wall outlet, electric streetcar, automobiles powered by gasoline, steam, and electricity, along with the tabletop stove, coffeemaker, bread machine, dishwasher and other wonders were displayed at the fair to amuse visitors and encourage new discoveries. Though they existed before the fair, the ice cream cone, cotton candy, iced tea, the hot dog, and hamburgers were greatly popularized there.

Problem. Find the opening year of the Louisiana Purchase Exposition if it is in the English language satisfies the following hints.

A. The first digit is the only number that is spelt with letters arranged in descending order.
B. The number of letters in the spelling of the second digit and the second digit itself are two consequent squares.
C. The third digit is not represented in Roman numerals.
D. The fourth digit is the only number that is spelt with the same number of letters as the number itself.

Answer. 1904

Yule Lads

Like many countries around the globe, Iceland celebrates Christmas with a joyful feast and gifts to loved ones. Kids anxiously wait for Father Christmas or Santa Claus. Icelandic children are also fortunate to have several Yule Lads. Each of them is known by name and personality. One of the Yule Lads visits each house one night before Christmas and puts a nice treat in a good kid's shoe left on the windowsill. Bad kids get their shoes filled with rotting potatoes. The Yule Lads used to be much creepier and other horrifying creatures could appear on Christmas Eve to scare kids. Grýla could come down from the mountains, and a giant blood-thirsty black Christmas Cat could prowl around the country and punish bad kids. Such monster stories were so chilling that parents were banned from telling them to their kids in 1746.

Problem. The number of Yule Lads can be presented as a sum of two-digit and one-digit numbers in two different ways. Both sums use the same set of three digits arranged in different ways. The product of the three digits is a prime number and their sum is a square number. Moreover, the word presentations of the two sums are *anagrams* of each other. How many Yule Lads do Icelandic children have?

Note. An *anagram* is a word or phrase formed by rearranging the letters of another word or phrase. For instance, *heart* and *earth* are anagrams formed from the same letters.

Answer. 13

Solutions

Countries Named Guinea. Let us look at the position of each number in the sequence, spell them, and count letters in the word:

1 – one – 3,
2 – two – 3,
3 – three – 5,
4 – four – 4,
5 – five – 4,
6 – six – 3,
7 – seven – 5,
8 – eight – 5,
9 – nine – 4,… .

It is easy to see that the obtained sequence becomes the given sequence 3, 3, 5, 4, X, 3, 5, 5, X, 3, 6, … and both missing numbers are 4s.

Equatorial Guinea, Guinea and Guinea-Bissau in Africa, and Papua New Guinea (German New Guinea) in Oceania and Asia are four countries with the word *Guinea* in their name. After gaining independence, French Guinea became Guinea, Spanish Guinea became Equatorial Guinea, and Portuguese Guinea became Guinea-Bissau.

Equatorial Guinea. The first step is to recover the missing numbers in this 27-term sequence. A close look at the sequence leads to the hypothesis that it is a combination of three sequences:

1, 2, 3, 4, …, 9; 0, 1, 2, 3, 4, …, 8; and 9, 8, 7, …, 1.

Then some missing numbers can be added to the original sequence as

1, 0, 9, 2, 1, 8, 3, 2, 7, 4, 3, 6, 5, 4, 5, 6, 5, 4, …, 9, 8, 1.

Number 5 appears at the 13th, 15th, and 17th positions.

Equatorial Guinea consists of Río Muni on continental Africa and five volcanic islands.

The Largest Gas Station. Only two products of five consequent numbers, $1{\cdot}2{\cdot}3{\cdot}4{\cdot}5 = 120$ and $2{\cdot}3{\cdot}4{\cdot}5{\cdot}6 = 720$, are three-digit numbers with 120 as the lowest.

Buc-ee's in New Braunfels, Texas, has 120 gas pumps and 1000 parking spots that leads to the world's record for the largest gas station.

Mont-Saint-Michel. Let us find the difference between two numbers described by Hint C:

$1{\cdot}2{\cdot}3 = 6$ – does not work because 6 is one-digit number;

$2{\cdot}3{\cdot}4$ – does not work because 2 and 4 are not co-primes;

$3{\cdot}4{\cdot}5 = 60$;

$4{\cdot}5{\cdot}6 = 120$ – does not work because it is a three-digit number.

Thus, only 60 satisfies the required condition of Hint C. Denoting the first, second, third, and forth digits of the UNESCO year and using 2 as the lowest prime given by Hint A as $a, b, c, c + 2$, we get $10c + c + 2 - 10a - b = 60$ by Hint C. Following Hint B, $10c + c + 2 = 11c + 2$, is prime, that is, c can take only odd values, and $11c + 2$ is greater than 60 by Hint C. Taking $c = 7$, we get 79 as the second two-digit number, $c = 5$ produces $57 < 60$, $c = 9$ has 111 that is a three-digit number. Thus, the UNESCO year is 1979.

Let us use Hint D to answer the second question. Taking tens digit 7, the difference between hundreds and units $9 - 9 = 0$, and the difference between hundreds and thousands units $9 - 1 = 8$ of the UNESCO year, we get 708.

Mont-Saint-Michel was listed as a UNESCO World Heritage Site in 1979. It was built following the dream of Aubert Bishop of Avranches he had in 708.

Coral Reefs. Only the squares of 1, 2, and 3 produce one-digit squares 1, 4, and 9 (Hint B). The difference between the square roots of the fourth and first numbers, $\sqrt{9} - \sqrt{1}$, is 2 (Hint C). Hence, the number is 1429.

The Great Barrier Reef spreads over 1429 miles taking approximately 133,000 square miles.

Three Geniuses. Only two one-digit numbers, 1 and 8, are cubes, and only three one-digit numbers, 1, 4, and 9, are squares. Just one of them, namely, 1, is a square and a cube. Then the first two digits of the year are 1 and then 8. Finally, let us test 4 for the last digit, then the third digit is $\sqrt[4]{1} \cdot \sqrt[3]{8} \cdot \sqrt[2]{4} =$ 4, which coincides with the last digit and cannot be used because of Hint D. Trying 9 for the last digit leads to the third digit of $\sqrt[4]{1} \cdot \sqrt[3]{8} \cdot \sqrt[2]{9} = 6$, and the year of 1869.

Indian Leader Mohandas Karamchand Gandhi, French artist Henri Émile Benoît Matisse, and Danish engineer Valdemar Poulsen were born in 1869.

The Braille System. Following the problem, a system of three linear and nonlinear equations in three variables can be constructed. It is challenging to solve it. Therefore, it would be better to analyze the problem first.

The first digit of a two-digit number squared to have another two-digit number is 0. Indeed, even $10^2 = 100$, which is a three-digit number. Hence, the third digit of the year is 0 (Hint C).

Let x and y be digits of the first two-digit number $10x + y$, and u be the second number. Then from Hints A and B we get

$$\begin{cases} 10x + y = 2u \\ x + y = 0 + u \end{cases}.$$

Subtracting the second equation from the first one we get $9x = u$, which is possible only if $x = 1$ and $u = 9$, because x, y, and u are digits. Then, $y = 8$, and the year is 1809.

Lous Braille was born in Paris in 1809. He published the first book *Method of Writing Words, Music, and Plain Songs by Means of Dots*, at age 20.

NATO. Let a, b, c, and $d, abcd \neq 0$, be the digits of year. Following the problem, we can write down the system

$$\begin{cases} \sqrt{b} = c - a \\ \sqrt{d} = \sqrt{c} + \sqrt{a} \end{cases}.$$

This system consists of two nonlinear equations in four variables. Various approaches can be employed to tackle the challenges of its solution. Below is one of them.

Since all digits of the year are nonnegative integers and their square roots are involved, then they are one-digit square numbers 0, 1, 4, 9. The first suggested number of 0 is excluded because the product of all digits of the year is positive and cannot be zero, $abcd \neq 0$. Let us prepare a table that outline all possible options for $\sqrt{b}$ and b (Table 5.1).

The only possible option is $c = 4$, $a = 1$, $b = 9$. Then the second equation leads to $d = 9$ and the year of 1949.

The same results are obtained if a table outlines all possible options for d.

The number of founded countries is 1 less than the sum of the tens and units digits of the year, that is, $4 + 9 - 1 = 12$. The number of NATO countries as of 2022 is 3 times the sum of the thousands and units digits of the year, that is, $3 \cdot (1 + 9) = 30$. The number of non-European countries is the square root of the tens digits of the year, that is $\sqrt{c} = 2$.

The intergovernmental military alliance NATO between 12 founding member states was signed in Washington, DC, on April 4, 1949. There are

TABLE 5.1

Possible options for hundreds digit $\sqrt{b}$, b

c \ a	1	4	9
1	0, –	–3, –	–8, –
4	3, 9	0, –	–5, –
9	8, –	5, –	0, –

28 European and two North American (the USA and Canada) countries as of 2022.

The World's Fair. The first digit is 1 because *one* is spelt with letters arranged in descending order as requested in Hint A. Hint B gives a clue for $9 = 3^2$ and its spelling *nine* with $4 = 2^2$ letters. 0 is not represented in Roman numerals, which following Hint C, gives the third digit. Finally, *four* has four letters that, according to Hint D, provides the fourth digit.

The Louisiana Purchase Exposition was an international exposition held in St. Louis, Missouri, United States, from April 30 to December 1, 1904.

Yule Lads. Let a, b, and c be three digits and $n = abc$ and $m = a + b + c$ be their product and sum. The product n of three numbers can be prime only if two of them are 1 and the third number is prime. Then $a = 1$ and $b = 1$. There are three different sums of two-digit and one-digit numbers: $11 + c$, $(10c + 1) + 1$, $(10 + c) + 1$. The first and the second sums $11 + c = (10c + 1) + 1$ are equal if $c = 10/9$, which is not integer. The first and the third sums are equal for any c, though only $c = 2$ and $c = 7$ lead to the sum $m = 1 + 1 + c$ presented as square numbers 4 and 9 correspondingly.

From the spelling of the two sums $11 + 2 = 12 + 1 = 13$ and $11 + 7 = 17 + 1 = 18$ as *eleven plus two is thirteen* and *twelve plus one is thirteen* and *eleven plus seven is eighteen* and *seventeen plus one is eighteen*, only the first phrase consists of anagrams. Moreover, it has 13 letters.

Icelandic kids have 13 Yule Lads that visit them on the 13 nights before Christmas. Following the poem written in 1932 by Jóhannes úr Kötlum, their names are Sheep-Cote Clod, Gully Gawk, Stubby, Spoon Licker, Pot Scraper, Bowl Licker, Door Slammer, Skyr Gobbler, Sausage Swiper, Window Peeper, Door Sniffer, Meat Hook, Candle Beggar.

Sources

www.smithsonianmag.com/smart-news/meet-the-thirteen-yule-lads-icelands-own-mischievous-santa-clauses-180948162/

www.worldatlas.com/articles/why-do-so-many-country-names-use-the-word-guinea,

www.economist.com/the-economist-explains/2017/09/12/why-the-world-has-so-many-guineas

www.factsinstitute.com/countries/facts-about-equatorial-guinea/

en.wikipedia.org/wiki/Mont-Saint-Michel

oceanservice.noaa.gov/facts/gbrlargeststructure.html, en.wikipedia.org/wiki/Coral_reef

en.wikipedia.org/wiki/Henri_Matisse

www.india.com/viral/mahatma-gandhi-6-fun-facts-you-didnt-know-about-him-9708/

www.britannica.com/facts/Mahatma-Gandhi

en.wikipedia.org/wiki/NATO

www.nato.int/

en.wikipedia.org/wiki/Louisiana_Purchase_Exposition

www.kcur.org/history/2021-09-14/st-louis-worlds-fair

source.wustl.edu/2004/04/xrays-fax-machines-and-ice-cream-cones-debut

whizz.com/en-us/blog/20-cool-facts-about-math

www.smithsonianmag.com/smart-news/meet-the-thirteen-yule-lads-icelands-own-mischievous-santa-clauses-180948162/

5.2 Number Puzzles

Math is Twice the Fun

You will be surprised to find that the distances between different cities around the world are the same.

Problem. Find the direct distances in miles between Aba (Nigeria) and Kinshasa (Zaire), between Ahmedabad and Chennai in India, and between Allahabad (India) and Peshawar (Pakistan), which are the same and are presented by the multiplicand (FUN). The direct distances between Amman (Jordan) and Milano (Italy), between Antalya (Turkey) and Sanaa (Yemen), and between Aurangabad (India) and Isfahan (Iran) are also the same and are hidden in the product (MATH) of the multiplication. To find these distances decode the word puzzle replacing letters with digits.

$$\begin{array}{r} \text{FUN} \\ \underline{\cdot \quad 2} \\ \text{MATH} \end{array}$$

There is a one-to-one correspondence between letters and digits. Digits from 0 to 9 are taken except for all one-digit multiples of one number.

Answer. 852 mi, 1704 mi

Roots are Bliss

The Assyrian New Year celebrates the rebirth of nature in spring. This day was marked as the first day of the year in the Babylonian calendar and was the biggest festival in the ancient Assyrian and Babylonian Empires that lasted 12 days. It is still widely celebrated in many countries around the globe.

Asteroids have brought lots of mystery, been used in prophecies, and taken human minds. Scientists have identified and classified more than millions of them.

Problem. Find the Assyrian New Year that started in spring of 2022 decoded by the addend ROOT. Discover what numbers are assigned to asteroids discovered at La Silla Observatory in 1988 (given by the addend ROOT) and in 1992 (shown by the sum BLISS).

Find the distances between several cities around the world if the addend ROOT provides the direct distance in miles between Anyang (China) and Montréal (Canada), between Belgrade (Serbia) and Santa Cruz de la Sierra (Bolivia), and between Brasília (Brazil) and Marrakesh (Morocco) and the total BLISS states the direct distance in kilometers between Adelaide (Australia) and Donetsk (Ukraine), between Gorakhpur (India) and Chihuahua (Mexico), and between Guatemala City (Guatemala) and Pretoria (South Africa).

Replace letters with digits and decode the following word puzzle:

$$\begin{array}{r} \text{ROOT} \\ +\,\text{ROOT} \\ \hline \text{BLISS} \end{array}$$

There is a unique correspondence between letters and digits.

Answer. 6772, 13,544

The Tallest Statue

The world's tallest statue, the Statue of Unity, is on a bank of the Narmada River in India. It is 182 meters or 597 feet tall if measured without the base! The statue was open in October 2018 in honor of the first deputy prime minister and home minister of independent India, Vallabhbhai Patel (1875–1950). He was a supporter and follower of legendary Mahatma Gandhi (1869–1948), who initiated nonviolent resistance while fighting for India's independence from British rule.

Problem. Find the height of the world's tallest Statue of Unity with its base if it is decoded as a multiplier in the following puzzle. The puzzle shows all 0s, 1s, and 8s appeared in the multiplication but hides all other digits. Present the result in meters and feet.

```
    xx0
    x8x
   ----
   1x80
   1xx0
  1x80
 ------
 188880
```

Answer. 240 m or 787 ft

Amazing Numbers

How many times can you fold a piece of paper? As many as you want to? No! Wrong! You cannot do this forever. But if you could, then the pile would reach the Moon. We use a quantitative description to say the number of times it is possible to fold a piece of paper.

Numbers are everywhere around us. It is difficult to imagine our life without numbers. We use them in our everyday routine, poetry, painting, counting, and measuring. Humanity have stepped far from tally and first counting means to modern computers. But… numbers remain. The names of numbers came to us from antiquity, then evolved and adapted to its current pronunciations in different languages, though similarities remained. Check, for instance, English *six*, Hebrew *shesh*, Spanish *seis*, and Russian *shest'*. The earliest forms of 11 and 12 are associated with Germanic *liban* meaning *leave*, which shows that *one is left* and *two is left* after counting to ten. Actually, *twelve* is the last number to be written as a word. The following 13 is the first number to be written using digits.

By the way, counting is not exclusive to humans. Animals can also count!

Problem. Decoding the puzzle in Figure 5.1 below, find

- the number that stands for hundred in Old Norse represented by the quotient,
- the maximum number of times paper can be folded presented by the divisor,
- the number of times Shakespeare mentioned *mathematics* in his plays given by the remainder.

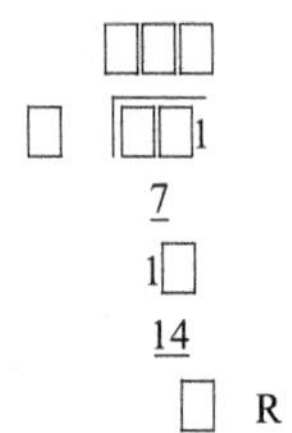

FIGURE 5.1

Can you name these plays by Shakespeare?

Answer: 120, 7, 1

Tug of War

While playing a tug of war, two teams tug opposite ends of a rope pulling it toward themselves. Though the origin of the game is uncertain, there are archaeological evidence and legends that tug of war was practiced in ancient Egypt, Greece, India, Cambodia, Papua New Guinea, China, and other places across the world. It was used during ancient ceremonies, rituals, competitions, and training soldiers. Tug of war has religious significance in Japan, Korea, and Myanmar. Different variations of this popular game have appeared over time. Two teams lying down on boards compete in New Zealand. Two teams battle on platforms suspended over a pool of water in Peru. Teams attempt to row towards each other on dragon boats in Poland and compete across the Mississippi River during Tug Fest in Illinois in USA. Tug of war is included in the World Games and was included in several Olympic Games. Cambodia, Philippines, Vietnam, and Korea registered tug of war as a UNESCO Intangible Cultural Heritage site in 2015.

Problem. How long are tug of war ropes if their lengths in feet available on Amazon are 1/4, 1/2, 1/3, and 1 of the longest length determined by the largest value among $2 \otimes 2, 3 \otimes 3, 4 \otimes 4$, and $5 \otimes 5$? The operation $a \otimes b$ is defined as

$$
\begin{array}{lll}
2 \otimes 3 = 20 & & 4 \otimes 5 = 40 \\
2 \otimes 4 = 20 & 3 \otimes 4 = 30 & 4 \otimes 6 = 40 \\
2 \otimes 5 = 20 & 3 \otimes 5 = 30 & \\
2 \otimes 6 = 20, & 3 \otimes 6 = 30, & 5 \otimes 6 = 50
\end{array}
$$

Note, that $a \otimes b \neq b \otimes a$. In particular, $3 \otimes 2 = 110$ but $2 \otimes 3 = 20$.

Answer. 25 ft, 50 ft, 75 ft, 100 ft

The Country Mirrors a River

The smallest country in mainland Africa, the Gambia, mirrors the shape of the of River Gambia, after which the country was named. The river and surrounding forest are a paradise for diverse wildlife. At its widest width, the Gambia is only 30 miles wide. Its territory extends almost 300 miles from the Atlantic coast into the interior of Africa. The country is surrounded by Senegal except for its western 50 miles of sunny sandy beaches on the Atlantic Ocean. Tourism and agriculture are its major sources of income.

Nine different tribes that speak their languages call this former British colony their home. Their warmest hospitality gives the country the nickname of *the smiling coast of Africa*. It is interesting that Gambians use marbles during elections and that the Gambia like the Bahamas should be referred to with the article *the*.

Problem. Find the narrowest width of the Gambia if this number in miles is presented in the outlined cells of the Mathdoku below (see Figure 5.2) after placing the numbers 1 to 9 in nine empty squares to have correct arithmetic operations:

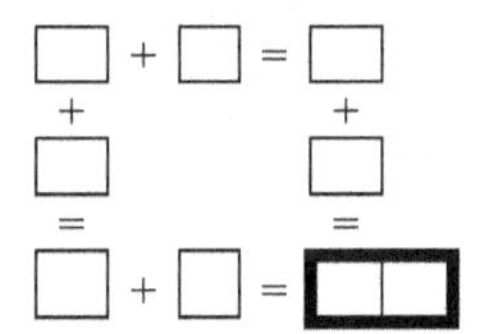

FIGURE 5.2

Answer. 15 mi

The Lone Star State

Texas is the second most populous and largest state in the United States. Its name means *friendly* in a Native American language. Native Americans lived in the area for thousands of years before the Spanish arrival in 1519. Mexico's war of independence pushed Spain away in 1821 and then, considering this land is not suitable for survival and habitation, 300 Anglo-American families were allowed to settle in Texas. A later fight with the Mexicans led to the formation of a country, called the Republic of Texas, until it agreed to join the United States as its 28th state and it is the only state to enter it by treaty, not by territorial annexation. Texas has no more right to secede than any other state but is allowed to be divided into up to five states. The state's nickname the *Lone Star State* comes from the time when it was independent with its single star flag that has remained up to now.

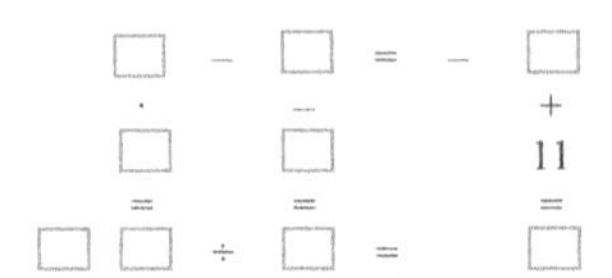

FIGURE 5.3

Texas is larger than any country in Western Europe. Its State Capitol is taller and larger than the US Capitol in Washington. Texas has the world leading NASA, a leading medical center to cure diseases, and is an oil producer. It pioneered in opening the first convenience store, invented the frozen margarita machine, and others. Two US presidents, Dwight D. Eisenhower and Lyndon B. Johnson, were born in Texas. Two other presidents, George H. W. Bush and George W. Bush, started their political careers there. Unfortunately, Texas has experienced the largest number of tornadoes and the deadliest natural disaster in the US.

Problem. What years did the Republic of Texas exist, if the last row of the Mathdoku shown in Figure 5.3 gives the year when the Republic of Texas gained independence from Mexico and the first number in the second row shows the number of years Texas was independent before joining the United States?

Answer. 1836–1845

The Sahara Desert

The People's Democratic Republic of Algeria became the largest country by area in Africa and the Arab League after South Sudan seceded from the Republic of the Sudan in 2011. Since the Sahara Desert takes the large portion of Algeria, most of Algerians live closer to the Mediterranean coast. Having petrol as one of the biggest sectors in the economy, Algeria provides some of the cheapest petrol in the world.

Algeria has a long and vivid history. It used to be the powerful Numidia Kingdom in Antiquity and its citizens were called Numidians and Imazighen, which means *Free Men*. The country is proud of having archeological sites, medinas, mountain villages, and fortresses that appear in the prestigious UNESCO World Heritage List. The original 1932 *Tarzan* movie was partially filmed in Algeria.

Problem. What portion of the territory of Algeria is covered by the Sahara Desert, if this number is hidden in the following puzzle shown in Figure 5.4?

Answer. 4/5

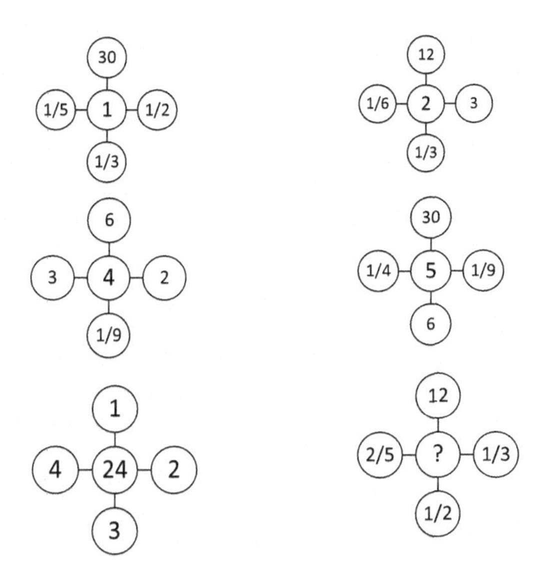

FIGURE 5.4

Criminals on Banknotes

Historic events, prominent citizens, and iconic structures are usually depictured on a country's currency. Australia is the only nation to ever feature criminals on its banknotes, some of them were convicted, others should have been. Indeed, Francis Greenway (1777–1837) was convicted of forging an architecture contract and was transported to New South Wales with a 14-year sentence. After submitting a breathtaking job application, which was accepted, he built town squares and beautiful buildings in Sydney that led to his appointment as the Civil Architect. An adventurous orphan Mary Reibey (1777–1855) was sent to Sydney at 13 with a seven-year sentence for stealing a horse but later became one of the wealthiest people after wisely managing her husband's business upon his death. These people take an opportunity to better themselves and set up a great example of rehabilitation. Their lives give a hope to others and show that given a second chance, convicts could prosper and build a new respectable future for themselves.

Francis Greenway is depicted on the a 1966 ten-dollar banknote but not on the new one issued in 1988, while Mary Reibey is featured on the current $20 banknote.

Problem. Find the number of people who flouted the law but later appeared on one of Australia's banknotes if this number is denoted by x in the following puzzle shown in Figure 5.5. Name them.

Answer. 3

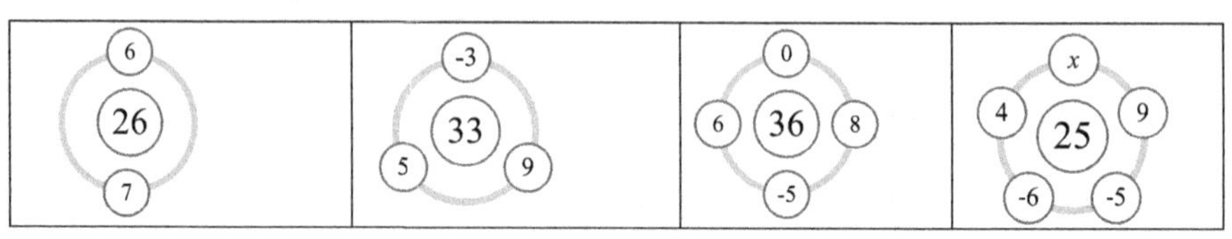

FIGURE 5.5

Units of Measurement

What do English scientist Isaac Newton (died in 1727), Swedish astronomer Anders Celsius (1701–1744), Italian physicist Alessandro Volta (1745–1827), and Scottish-Canadian inventor Alexander Graham Bell (1847–1922) have in common? These outstanding scientists lived during different time periods in different countries and had never met, but all of them are legendary for their inventions. Celsius suggested a temperature scale used worldwide where water freezes at 0° degrees and boils at 100°. Using his tongue to measure differences in electrical potential, Volta discovered methane and invented the first electric battery. Working with deaf people, Bell experimented with sound and built an instrument for transmitting sound via electricity giving the birth to the first phone. The contribution of Issac Newton to development of science is incredible. He revealed that white light is composed of different

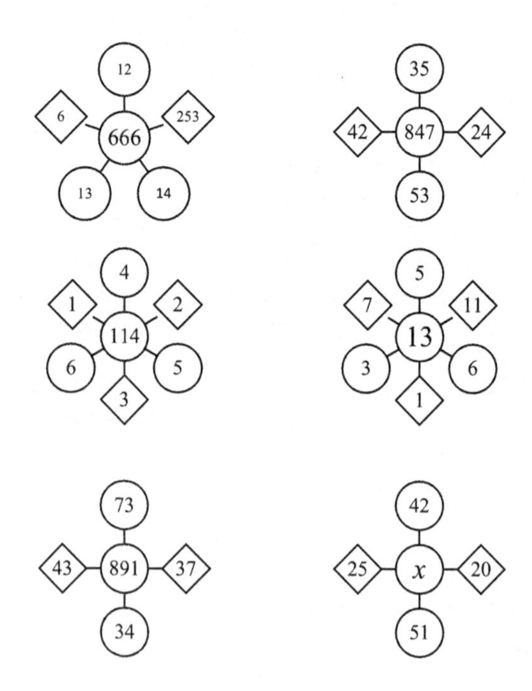

FIGURE 5.6

colors, discovered gravity and the three laws of motion, and was one of two founders of calculus in addition to his other numerous achievements. These greatest inventors have their names immortalized in international units named after them. Indeed, force is measured in Newton (N), sound magnitude is given in Bell (B), Celsius (°C) is used to measure temperature. Volts (V) are used to measure electric potential.

Problem. When was Isaac Newton born if this year is decoded by x in the following puzzle shown in Figure 5.6?

Answer. 1642

Solutions

Math Is Twice the Fun. There are different ways to decode the puzzle.

There are ten digits. FUN and MATH need seven digits to be replaced. Thus, three digits should be taken away. These digits are all one-digit multiples of one number, which, following the statement below, is 3.

- 1 has 9 multiples,
- 2 has 4 multiples, 2, 4, 6, and 8,
- 3 has 3 multiples, 3, 6, and 9,
- 4 has 2 multiples, 4 and 8,
- 5, 6, 7, 8, 9 have only one one-digit multiple.

Hence, letters should be replaced with digits 0, 1, 2, 4, 5, 7, 8. As the leading digit M of the product and a one-digit number doubled, M = 1 and F ≥ 5.

N cannot be

- 1 because 1 has been already taken by M,
- 0 because then H = 0 (one digit cannot be taken by different letters),
- 3, 6, or 9 as multiples of 3,
- 8, because then H = 6, that has been eliminated as a multiple of 3.

Thus, N can be 2, 4, 5, or 7.

Continuing similar reasoning and providing eliminations of not suitable options, we get F = 8, U = 5, N = 2, A = 7, T = 0, and H = 4.

There are the 852 miles (1370 km) of direct distance between cities in Nigeria and Zaire, between Ahmedabad and Chennai in India, between Allahābād (India) and Peshawar (Pakistan). There is a 1704-mile (2742 km) direct distance between Amman (Jordan) and Milano (Italy), between Antalya (Turkey) and Sanaa (Yemen), and between Aurangabad (India) and Isfahan (Iran).

Roots Are Bliss. There are different ways to decode the puzzle.

As B is the first digit of a number obtained from adding two one-digit numbers R + R, then $B = 1$ and $R \geq 5$.

As the sum of T and T, S is even but not 0 because then T will be also 0, which contradicts the condition that all digits should be different.

T can be any digit less than 4 because S is even and 1 cannot be carried out to the second S (tens S in BLISS) that would make S odd if $T > 4$. Hence, T can be 2, 3, or 4.

Two Os are added twice, but the second time the last digit of addition is I. Therefore, $R \geq 5$ (as before) and $I = S + 1$, because only 1 can be carried out after adding two one-digit numbers.

Possible options of related digit values for T, S, I, and O are presented in the table below.

T	S	I	O
2	4	5	7
3	6	7	8
4	8	9	9

The last row with $T = 4$ should be taken out because then I and O are the same.

Let us consider different values for R, namely, 5, 6, 7, 8 and 9. $R \neq 5$, because then $L = 1$ (but $B = 1$), $R \neq 7$, because then $I = 7$ or $O = 7$ in remained two lines of the table, $R \neq 8$, because then $L = 1$ (but $B = 1$). Continuing the process, we can show that $R \neq 8$ and $R \neq 9$. Then $R = 6$ and the remaining first line of the table gives values for other letters. The decoded puzzle is $6772 + 6772 = 13544$.

6772 is the Assyrian New Year that is celebrated in spring of 2022.

Asteroids with numbers 6772 and 13,544 were discovered in La Silla Observatory in 1988 and 1992 correspondingly.

The direct distance between Anyang (China) and Montréal (Canada), between Belgrade (Serbia) and Santa Cruz de la Sierra (Bolivia), and between Brasília (Brazil) and Marrakesh (Morocco) is 6772 miles or 10,898 km.

The direct distance between Adelaide (Australia) and Donetsk (Ukraine), between Gorakhpur (India) and Chihuahua (Mexico), and between Guatemala City (Guatemala) and Pretoria (South Africa) is 13,544 km or 8416 miles.

The Tallest Statue. There are multiple ways to decode the puzzle. To present one of them and make explanations easier, let us substitute crosses with letters:

```
   ab0
   c8d
 ------
   1e80
  1fg0
 1h80
 ------
 188880
```

From the column $1 + f + 8 = 8$, it follows $f = 9$ and 1 is carried out.
The number 19g0 is divisible by 8 without a remainder only if $g = 2$.
The quotient after 1920 is divided by 8 is 240. Thus, $a = 2$ and $b = 4$.
From the column $1 + h = 8$ and remembering carrying 1 follows $h = 6$.
From the column $e + g + 0 = 8$ or $e + 2 + 0 = 8$ follows $e = 6$.
The quotient after 1680 divided by 240 is 7. Thus, $c = 7$ and $d = 7$.

Hence, $240 \cdot 787 = 188{,}880$ provides two multipliers. Which one to take? Looking carefully at the multipliers and their relation, we can find that they represent the same height expressed in meters and feet.

The height of the Statue of Unity including the base is 240 meters or 787 feet.

Amazing Numbers. The puzzle can be solved in different ways. One of them is shown in Figure 5.7. Let us check the divisor, first digit of the quotient, and 7. Since 7 is prime, then only $7 = 1 \cdot 7$ is possible, and the divisor or first digit quotient can be either 1 or 7. Since 14 is a two-digit number, then the divisor is 7 and the first digit of the quotient is 1. Because of 14, the second digit of the quotient is 2 and the first two digits of the dividend are 8 and 4. Finally, 1 as the last digit of the dividend leads to the remainder of 1 and the last digit of the quotient of 0.
The *hundred* comes from an Old Norse *hundrath* meaning 120. A piece of paper can only be folded seven times. Shakespeare only mentioned *mathematics* in his play *The Taming of the Shrew*.

Tug of War. Noticing that the answer in $3 \otimes 2 = 110$uses only two digits from 0 to 1, and the answer in $2 \otimes 3 = 20$may use digits ranging from 0 to 2, we can assume that the operation $a \otimes b$is a multiplication where the second factor provides the base of a numeral system the result of the multiplication should be transformed to. Indeed, $3 \cdot 2 = 6$ in the decimal system but $110_2 = 1 \cdot 2^2 + 1 \cdot 2^1 + 0 \cdot 2^0 = 6_{10}$ in the binary system and $3 \otimes 2 = 110$. Similarly, $2 \cdot 3 = 6$ in the decimal system but $20_3 = 2 \cdot 3^1 + 2 \cdot 3^0 = 6_{10}$ in the system to base 3, and $2 \otimes 3 = 20$. Hence, the operation$a \otimes b$supports the initial idea that two numbers a and b are multiplied and then the product is represented as a number in base b. This hypothesis can be verified with the provided table to assure that it is true. Indeed, $2 \cdot 4 = 8$ that in base 4 is 20. Then $2 \otimes 2 = 100$ in base 2 or 4 in base 10,

```
     120
 7  √841
     7
     14
     14
      1  R
```

FIGURE 5.7

$3 \otimes 3$= 100 in base 3 or 9 in base 10, $4 \otimes 4$= 100 in base 4 or 16 in base 10, $5 \otimes 5$ = 100 in base 5 or 25 in base 10. That is, all$2 \otimes 2, 3 \otimes 3, 4 \otimes 4, 5 \otimes 5$result in 100.

The lengths of tug of war ropes available on Amason are 25 ft, 50 ft, 75 ft, 100 ft.

The Country Mirrors a River. Using the trial-and-error method we can get the following result shown in Figure 5.8.
At its narrowest, the Gambia is 15 miles or 25 km wide.

4 _	+	3	=	7
+				+
2 _				8
=				=
6 _	+	9	=	1 5

FIGURE 5.8

The Lone Star State. The problem has various ways that leads to the unique completion of the table.

	2	–	7	=	–	5
	·		–			+
	9		4		1	1
	=		=			=
1	8	÷	3	=		6

The Republic of Texas gained independence from Mexico in 1836. It existed as an inwdependent country for 9 years until it agreed to join the United States in 1845.

The Sahara Desert. It is easy to notice that the center number is the product of all outer numbers. Indeed, $1 \cdot 2 \cdot 3 \cdot 4 = 24$, $30 \cdot 1/2 \cdot 1/3 \cdot 1/5 = 1$, and so on. Thus, $12 \cdot 1/3 \cdot 1/2 \cdot 2/5 = 4/5$.

Over four-fifths of Algerian territory is covered by the Sahara Desert.

Criminals on Banknotes. A close look at the puzzles gives a hint that the center number is the sum of numbers in the circles multiplied by the number of circles. Indeed, $(6 + 7) \cdot 2 = 26$, $(6 + 0 + 8 - 5) \cdot 4 = 36$. Then from $(4 + x + 9 - 5 - 6) \cdot 5 = 25$ follows $x = 3$.

Three criminals, Francis Greenway, John Macarthur, and Mary Reibey, can be found on Australia's currency. John Macarthur appeared on \$2 note of Australia's currency before it was replaced with a coin.

Units of Measurement. A close look at the puzzles tells us that the middle number is the difference between the product of numbers in circles and the product of numbers in rombas. Indeed, $4{\cdot}5{\cdot}6 - 1{\cdot}2{\cdot}3 = 114$ and $73{\cdot}34 - 43{\cdot}37 = 891$. Hence, $x = 42{\cdot}51 - 20{\cdot}21 = 1642$.

English scientist Isaac Newton was born in 1642.

Sources

www.thecollector.com/three-criminals-that-surprisingly-appeared-on-australias-currency/

www.arabamerica.com/12-fun-facts-about-algeria-you-probably-didnt-know/

en.wikipedia.org/wiki/Algeria

www.history.com/news/9-things-you-may-not-know-about-texas

thefactfile.org/texas-facts/2/

medium.com/@hb20007/top-10-countries-with-the-most-beautiful-shapes-on-the-map-f90cb8ebcd62

www.gambia.co.uk/blog/10-interesting-facts-about-the-gambia

www.mathnasium.com/eagan/news/some-amazing-math-facts

en.wikipedia.org/wiki/Statue_of_Unity

www.wincalendar.com/Assyrian-New-Year

en.wikipedia.org/wiki/Tug_of_war

en.wikipedia.org/wiki/List_of_scientists_whose_names_are_used_as_units

5.3 Grid Puzzles

The first problem, *Heart Sudoku,* gives a clue to solve several of the next problems of this section.

Heart Sudoku

The development of Sudoku goes back to a game, *Latin Square,* proposed by Swiss mathematician Leonhard Euler in 1783. It was first appeared in French newspapers a century later. American Howard Garns suggested and published the *Number Place* puzzle in 1979. Japanese Maki Kaji spotted the puzzle, realized its advantages, and published it in Japan in 1984 calling it *Sudoku,* which is short for a long Japanese expression meaning *the digits are limited to one occurrence.* Judge Wayne Gould from New Zealand came across Sudoku and brought it back to the Western world where they were first

		1				3		
	2	3	5		6	1	8	
4	5		1		7		2	6
5				1				9
2	3						6	1
	8	7				4	3	
		2	3		5	6		
			9	7	2			
				8				

FIGURE 5.9

published in British and then American newspapers in 2004. Sudoku puzzles have conquered the word extremely quickly.

Fill in the *Heart Sudoku* shown in Figure 5.9 and solve the following four problems to learn some amazing facts about official and spoken languages in countries around the globe.

Note. A 9×9 Sudoku square must be filled in with digits from 1 to 9 that are not repeated in any horizontal line, any vertical line, or any 3×3 square-marked grid.

Languages around the Globe

How many different languages do people speak? How many languages do citizens of one country speak? It is not easy to answer. Historical events have shaped and re-shaped our world. To recognize its multicultural population and respect cultural identities, the constitution of many countries acknowledges several official languages. Let us take a look at just three countries from three different continents.

The Republic of Singapore is a sovereign island city-state off the southern tip of the Malay Peninsula in Southeast Asia. Having the third largest population density in the world, Singapore is continuously reclaiming land from the sea to increase the area of its main island and small islands.

Switzerland is a Central European country with astonishing scenery of the Alps, beautiful lakes, and cities with medieval quarters. Swiss watches and chocolate are world famous.

The Republic of Zimbabwe, often called the *Jewel of Africa*, has one of the richest heritage among Southeast African countries.

Problem. Switzerland and Singapore have the same number of official languages. This number is the lowest composite number on the main diagonal of the Heart Sudoku puzzle, while this number squared is the number of official languages in Zimbabwe. How many official languages are in Switzerland, Singapore, and Zimbabwe?

The square root of the largest composite number on the mail diagonal is the number of languages official in over n counties, where n is the sum of the largest and smallest composite numbers. Find these numbers.

Answer. 4, 4, 16, 3, 13

Official Languages in Palau

The Republic of Palau is an archipelago of over 500 islands in Micronesia in the western Pacific Ocean. The area of this sovereign nation is only around 466 km^2, almost twice smaller than New York in the US. Palau's lakes are filled with jellyfish and saltwater crocodiles. The Palauan delicious food is a soup made from fruit bats, coconut milk, ginger, and spices. It is interesting that this paradise country does not have a military.

Problem. How many official languages does Palau have if this number is the prime number that appears more frequently on the main diagonal of the Heart Sudoku?

Answer. 2

Official Languages in Australia

The area of the Commonwealth of Australia is the largest in Oceania and is among the largest in the world. The name of the country came from Latin *Terra Australis* or *southern land* given to an imaginary continent in the Southern Hemisphere. Although inhabited for over 65,000 years, the naval exploration of Australia by Dutch started in the 17th century. Its eastern part was claimed by Great Britain in 1770. The European population grew fast making English the dominant and de facto national language, despite over 400 Aboriginal languages spoken there.

Problem. The number of official languages in Australia can be expressed as the difference between two numbers symmetric about the main diagonal of the Heart Sudoku. One number is located at the intersection of Row and Column presented as

$$\frac{(\sqrt{a}-3)(a\sqrt{a}+27)}{(a-9)(a-3\sqrt{a}+9)} \text{ and } \frac{\sqrt{(a-5)^2-a^2+2\cdot 5a}}{\sqrt[3]{(a-5)^3-a^2(a-3\cdot 5)-5^2(3a-5)+5^0}}$$

correspondingly. How many official languages are there in Australia?

Answer. 0

Official Languages in the USA and Mexico

The United States of America and Mexico are neighbors. Their 1954-mi border from the Gulf of Mexico to the Pacific Ocean is only marked by a sign in some places. Both countries are members of Organization of American States and the United Nations. Since the late 19th century, the two countries have had close diplomatic and economic ties. They also have the same number of official languages. Can you name them? Yes, you know, English in the USA and Spanish in Mexico. Check your responses after solving the problem.

Problem. Determine the number of official languages in Mexico and the USA if it is described by two numbers in the Heart Sudoku puzzle after subtracting the second number squared from the first number. The two numbers share the same column defined by the resulting exponent of a in the following expression

$$\frac{\sqrt{ab^{-3}}\cdot\sqrt[3]{a^7b^4}\cdot\sqrt[4]{(ab)^{-3}}}{\sqrt{(a^3b^7)^{-1}}\cdot\sqrt[3]{ab}\cdot\sqrt[4]{a^{-4}(ab)^{-7}}}.$$

The first number is a diagonal number, and the second number is located on the row defined by the resulting exponent of b in the expression.

Answer. 0

Inventions and Complaints

Penicillin, Viagra, anesthesia, microwaves, and many other inventions have been discovered by accident. History is also full of examples of inventions caused by complaints and bad reviews. American playwright Joseph Kesselring received negative reviews on his fiction with recommendations to rewrite it as a comedy. He followed the advice and presented his by far most successful black comedy, *Arsenic and Old Lace*. The play is so funny, that it still plays in theaters worldwide and has been the plot of several movies.

Another example is related to one of our favorite foods. While serving French fries, African American chef George Crum received complaints that the slices were too thick. To please customers, he made the potato slices thinner and thinner and finally offered the Saratoga Chips that we enjoy today.

Problem. Find the year when chips were offered by chef George Crum if this year is shown in bold cells in the following Kakuro shown in Figure 5.10.

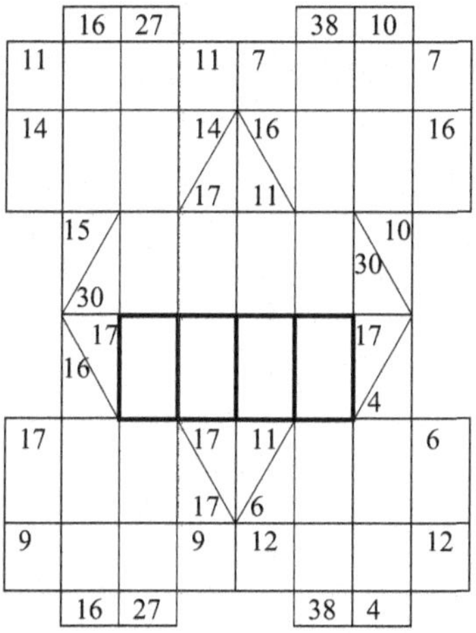

FIGURE 5.10

Note. Kakuro, translated from Japanese as *addition cross*, is a logic crosswords-like puzzle, where digits from 1 to 9 should be placed to make the sum indicated by numbers written at the top and bottom, to the left and right. A digit can appear only once inside each sum group.

Answer. 1853

Discovery of Insulin

The disease *diabetes mellitus* has been known since ancient times and traced back to old Egyptian, Indian and Chinese texts. The 1st century text documented the signs of the illness. Various attempts were made to cure this disease. English Thomas Willis detected sugar-like ingredients in patients' urine in the 17th century. Noticing that fasting and dietary restriction may improve symptoms of diabetes, British surgeon John Rollo successfully treated his patients in 1706. The modern father of diabetology, French pharmacist Apollinaire Bouchardat, also advised sugar-free diets in the 19th century. The real progress, however, was made with detecting a chemical that was missing from the pancreas in people with diabetes made by an English physiologist, Sir Edward Albert Sharpey-Shafer, in 1919. He called it *insulin* from the Latin word *insula*, meaning *island*. The revolutionary discovery of isolating the hormone insulin was made by young Canadian surgeon Sir Frederick G. Banting and his assistants Charles H. Best and J. J. R. Macleod. The insulin was later purified by James B. Collip. One year later the first injection of insulin treated a boy dying from diabetes. For their genuine breakthrough Banting and Macleod received the Nobel Prize in Medicine, which they shared with Best and Collip.

Problem. When was the hormone insulin isolated if two cells in bold of the magic square jointing together contain this year (see Figure 5.11)?

	24	1	8	15
		7	14	16
4			20	22
10	12	**[ ]**	**[ ]**	3
11	18	25		

FIGURE 5.11

Note. The primary *magic square* consists of the same number n of rows and columns filled with numbers from 1 to n^2, such that the *magic total* or the sums of all entries of each row, column, and diagonal are the same. Different variations of the magic square can use different sets of numbers though retain main properties.

Answer. 1921

Tectonic Plates

The lithosphere or hard surface of our planet stands on a semi-molten layer of rocks and upper mantle. The rock density varies significantly that leads to the appearance of cracks and different plates. Tectonic plates move slowly but constantly causing various disasters and forming oceans, mountains, islands, and continents. Iceland appeared as the result of turbulent interactions between the Eurasian and North American tectonic plates. Japan sits on four major tectonic plates that bring frequent earthquakes, volcanoes, and trenches to the country. Earthquakes and tsunamis in Indonesia are also due to its position on three major tectonic plates. Mexico is one of the most seismically active countries because it sits atop three largest tectonic plates.

Problem. How many major tectonic plates does the Earth have if this number is hidden in the marked cells of the magic square shown in Figure 5.12? Name them.

6		14
	13	

FIGURE 5.12

Note. This variation of a *magic square* consists of n rows and columns, keeps the same sums of all entries in each row, column, and diagonal but is filled with any numbers, not necessarily from 1 to n^2.

Answer. 7

Metro in Turkey

Istanbul in Turkey is a transcontinental city located in both Europe and Asia. The city was founded as Byzantium in the 7th century BC by Greek settlers. During its amazingly rich history, Istanbul was named New Rome and Constantinople. Surprisingly, after serving as the capital city for the Roman/Byzantine, Latin, Byzantine, and Ottoman empires, Istanbul ceded this privilege to Ankara following the Turkish War of Independence in 1923. Istanbul is the largest city in Turkey and its economic and historic center. The Istanbul Metro was constructed to serve this most populous European city.

Problem. Find the opening year of the Istanbul Metro if

- this year can be presented with two numbers from the *Reversible Magic Square* (see Figure 5.13) written side by side and read upside-down,
- the first number is the entry at the intersection of the fourth row and second column,
- the second number is the entry at the intersection of the first row and fourth column,
- the sign ~ stands for different two-digit numbers with the same digits that can be also read upside-down,
- three pairs of symbols X, V, and S represent the related pairs of two-digit numbers upside-down to each other, like 18 and 81,
- there are no zeroes among digits in this magic square.

Note. The sums of all entries in each row, column, and diagonal of a magic square and its reversible version are the same. The numbers in the magic square cannot be repeated.

Answer. 1989

18	~	86	X
~	81	98	X
V	V	69	~
S	S	~	96

FIGURE 5.13

The Eiffel Tower

The Eiffel Tower is one of the world's most recognizable landmarks. Its replicas can be found around the world. The tower was built in two years and two months for the Universal Exhibition in Paris to celebrate the centennial of the French Revolution and display mechanical achievements of the country. Despite the initial intention to demolish it in 20 years, the Eiffel Tower was constructed fundamentally with a surprising secret apartment, post office, scientific laboratory, and military bunker. The names of 72 19th-century French scientists are engraved on the Eiffel Tower. The Eiffel tower has survived its destroying predictions and has become one of the symbols of Paris. This Iron Lady has been extremely fashionable, changing her look from a reddish-brown color to yellow, to yellow-brown, and, finally, to its current three shades of brown.

Problem. Find the opening year of the Eiffel Tower and when she chose her latest colors if these years are hidden in the *Reversible Magic Square* shown in Figure 5.14. The opening year of the Eiffel Tower is presented by two numbers located at the intersection of Row 1 and Column 1 and of Row 4 and Column 1 written side by side. The year when the Eiffel Tower chose her current color can be presented with two numbers located at the intersection of Row 2 and Column 4 and of Row 4 and Column 2 written side by side.

	99			
66				
			88	
		11		

FIGURE 5.14

Note. The *magic total sums* of all entries at each row, column, and diagonal of a magic square are the same. A *Reversible Magic Square* keeps the same magic total if read upside down. The sums of entries at the marked rectangles in the magic reversible square and its upside-down square are also equal to the magic total. All two-digit numbers are different. Each row and each column have all four digits, 1, 6, 8, and 9, as tens and ones.

Answer. 1889, 1968

Countries Rich with Lakes

Finland is named the land of a thousand lakes. Its forested landscape is scattered with water spots. Brazil has the highest volume of freshwater. Canada is the champion for having the world's greatest number of lakes that take 9% of its territory and account for over half of the world's natural lakes. Canadian lakes have enough water to flood the entire country to up to 2 meters. Its largest lakes are the Great Bear and Great Slave Lakes.

Problem. How many lakes with the area of at least 3 km^2 does Canada have if this number is represented by all digits of the second row of the Inky puzzle written side by side with digit 4 replaced by 7 (shown in Figure 5.15)?

6+		5/		5+
3/		8+	3-	
12x				20x
6+	3-			
		24x		

FIGURE 5.15

Note. Use digits from 1 to 5 to fill in the *Inky* or *KenKen* puzzle. Each area in bold, called a cage, contains a group of digits that can be combined in any order using the indicated mathematical sign to produce the result shown in the cage. The same digits can appear in one cage but neither in one row nor one column.

Answer. 31,752

Solutions

The Heart Sudoku

Sudoku solution is shown in Figure 5.16.

Languages around the Globe. The lowest and largest composite numbers on the main diagonal of the Sudoku puzzle are 4 and 9.

Switzerland has four official languages, German, French, Italian, and Romansh. Singapore also has four official languages, English, Mandarin Chinese, Malay, and Tamil, while Zimbabwe has 16 official languages with English, Shona, and Ndebele as the most common.

		1				3		
6	9		4	2	8		5	7
	2	3	5		6	1	8	
7				9				4
4	5		1		7		2	6
		8		3		9		
5				1				9
	6	4	8		3	2	7	
2	3						6	1
		9	7	5	4	8		
	8	7				4	3	
1			2	6	9			5
		2	3		5	6		
9	7			4			1	8
			9	7	2			
8	1	6				5	4	3
				8				
3	4	5	6		1	7	9	2

FIGURE 5.16

Over 13 counties have three official languages.

Official Languages in Palau. There are two prime numbers, 2 and 5, on the main diagonal of Sudoku. Prime 2 appears twice, while 5 appears ones.

Palau has two official languages, Palauan (from a Malayo-Polynesian group) and English.

Official Languages in Australia. Let us simplify the expressions using $a\sqrt{a} = (\sqrt{a})^3$and algebraic identities

$$(a-b)^2 = a^2 - 2ab + b^2,\ (a-b)^3 = a^3 - 3a^2b + 3ab^2 - b^3,\ a^3 + b^3 = (a+b)(a^2 - ab + b^2).$$

Then the row number becomes

$$\frac{(\sqrt{a}-3)(a\sqrt{a}+27)}{(a-9)(a-3\sqrt{a}+9)} = \frac{(\sqrt{a}-3)(\sqrt{a}+3)(a-3\sqrt{a}+9)}{(a-9)(a-3\sqrt{a}+9)} = 1;$$

and the column number turns to be

$$\frac{\sqrt{(a-5)^2 - a^2 + 2\cdot 5a}}{\sqrt[3]{(a-5)^3 - a^2(a-3\cdot 5) - 5^2(3a-5) + 5^0}}$$

$$= \frac{\sqrt{a^2 - 10a + 25 - a^2 + 2\cdot 5a}}{\sqrt[3]{a^3 - 15a^2 + 75a - 125 - a^3 + 3\cdot 5a^2 - 5^2\cdot 3a + 5^3 + 1}} = 5.$$

Number 2 is located at the intersection of Row 1 and Column 5. The second number symmetric about the main diagonal in Sudoku is at Row 5 and Column 1. It is 2 as well. Their difference is 0.

Surprisingly, Australia does not have any official language.

Official Languages in the USA and Mexico. Let us rewrite the provided expression

$$\frac{\sqrt{ab^{-3}}\cdot\sqrt[3]{a^7b^4}\cdot\sqrt[4]{(ab)^{-3}}}{\sqrt{(a^3b^7)^{-1}}\cdot\sqrt[3]{ab}\cdot\sqrt[4]{a^{-4}(ab)^{-7}}} \text{ as } \frac{a^{\frac{1}{2}}b^{-\frac{3}{2}}\cdot a^{\frac{7}{3}}b^{\frac{4}{3}}\cdot a^{-\frac{3}{4}}b^{-\frac{3}{4}}}{a^{-\frac{3}{2}}b^{-\frac{7}{2}}\cdot a^{\frac{1}{3}}b^{\frac{1}{3}}\cdot a^{-\frac{4}{4}}a^{-\frac{7}{4}}b^{-\frac{7}{4}}}$$

and combine exponents of a and b as

$$\frac{1}{2}+\frac{3}{2}+\frac{7}{3}-\frac{1}{3}+\frac{7}{4}+\frac{4}{4}-\frac{3}{4}=6 \text{ and } \frac{7}{2}-\frac{3}{2}+\frac{4}{3}-\frac{1}{3}+\frac{7}{4}-\frac{3}{4}=4,$$

correspondingly. The first number positioned at the intersection of Row 6 and Column 6 is 9, the second number is at the intersection of Row 4 and Column 6 is 3. Then $9-3^2=0$.

Surprised? Yes, both countries do not have any official languages.

The right of Americans to speak any language is protected by the US Constitution that also allows for states to declare their official languages. John Adams' bill to Congress in 1780 to make English the official language of the United States did not pass because of the fear of threatening individual liberty.

Mexico recognizes 68 national languages, including Nahuatl, Maya, and other indigenous languages. Spanish is used in most Mexican government proceedings and is spoken by most of its population. However, Mexico has no official language.

Inventions and Complaints. To solve the Kakuro, we have to find cells at the intersections of rows and columns that can be easily filled in. For example, the upper right column with 16 can be filled only with 9 and 7. But 9 cannot go first, because then 7 will go down and 14 will be written as 7 + 7, which contradicts the rule that numbers cannot be repeated. Therefore, 7 goes up leading to 7 + 4 = 11, then 9 goes down giving 9 + 5 = 14, and so on. The solution is shown in Figure 5.17.

Potato chips were first made by African American chief George Crum in 1853.

	16	27			38	10	
11	7	4	11	7	6	1	7
14	9	5	14 / 17	16 / 11	7	9	16
	15 / 30	7	9	6	8	10 / 30	
	17 / 16	**1**	**8**	**5**	**3**	17 / 4	
17	9	8	17 / 17	11 / 6	5	1	6
9	7	2	9	12	9	3	12
	16	27			38	4	

FIGURE 5.17

Discovery of Insulin. Let us find the *magic total* of the 5×5 magic square. There are 5·5 =25 consequent numbers from 1 to 25 that can be considered as an arithmetic sequence with the first term of 1 and the difference of 1.

The sum of its 25 terms is

$$S = \frac{a_1 + a_n}{2} n = \frac{1+25}{2} \cdot 25 = 325.$$

Since there are 5 rows (and columns), the magic sum is 65 = 325/5. Then entry a_{11} at the intersection of the first row and the first column is 65 – 24 – 1 – 8 – 15 = 17. Calculating a_{21} as 65 – 17 – 4 – 10 – 11 = 23, we can find a_{22} as 65 – 23 – 7 – 14 – 16 = 5, and so on until the resulting table shown in Figure 5.18 is completed.

17	24	1	8	15
23	5	7	14	16
4	6	13	20	22
10	12	**19**	**21**	3
11	18	25	2	9

FIGURE 5.18

Insulin was invented in 1921.

Tectonic Plates. To find the magic total, let us denote the central entry as a and the entries of the first row as x, y, and z, of the third row as u, v, and w.

x	y	z
	a	
u	v	w

Since the total sums of all entries in each row, column, and diagonal are the same, say S, we can write

$x + a + w = S \Rightarrow x = S - a - w; y + a + v = S \Rightarrow y = S - a - v;$

$u + a + z = S \Rightarrow z = S - a - u; u + v + w = S.$

Then $S = x + y + z = S - a - w + S - a - v + S - a - u = 3S - 3a - (u + v + w) = 3S - 3a - S \Rightarrow S = 3a.$

Using formulas above and given values $x = 6, a = 13, z = 14$, and $S = 3{\cdot}13 = 39$, we can get $w = 39 - 13 - 6 = 20$, $u = 39 - 13 - 14 = 12$, and $v = 39 - 12 - 20 = 7$.

6	19	14
21	13	5
12	7	20

Pacific Plate, North American Plate, Eurasian Plate, African Plate, Antarctic Plate, Indo-Australia Plate, South American Plate are seven major tectonic plates. There are many smaller tectonic plates.

Note. You can construct your own magic square following the rule below, where a, b, and c can be any numbers, for instance, the birthday of your friend.

$a - b$	$a + b - c$	$a + c$
$a + b - c$	a	$a - b - c$
$a - c$	$a - b + c$	$a + b$

Metro in Turkey. Adding entries of the main diagonal, $18 + 81 + 69 + 96 = 264$, the magic total of this Reversible Magic Square is 264, which is also the sum of all numbers at any diagonal, any row, or any column.

The problem can be solved in different ways.

There are four two-digit numbers with the same digits that can also be read upside down, 11, 66, 88, and 99. These numbers can take a cell denoted by ~.

From Column 3: $264 - 86 - 98 - 69 = 11$ that takes the cell at the intersection of Row 4 and Column 3.

From Row 2: $264 - 81 - 98 = 85$ for two entries in the second row. Because 11 has been taken, only one number less than 85, namely 66, is left. It occupies the cell at the intersection of Row 2 and Column 1.

Then $19 = 85 - 66$ is for the cell at the intersection of Row 2 and Column 4 and its upside-down number 61 for the cell at the intersection of Row 1 and Column 4.

From Column 4: $264 - 61 - 19 - 96 = 88$ is for the cell at the intersection of Row 3 and Column 4.

Finally, the remaining 99 takes a cell at the intersection of Row 1 and Column 2.

To find the entries for the left empty cells, the system of four linear equations in four variables can be constructed or close consideration and logical thinking may help to find these entries.

From Column 2: 264 – 99 – 81 = 84 that can be presented as 16 + 68. These numbers and their upside-down numbers 91 and 89 will take the remaining cells.

The number composed from entries at the intersection of Row 4 and Column 2 and Row 1 and Column 4 written side by side is 6861 that read upside down gives 1989.

18	99	86	61
66	81	98	19
91	16	69	88
89	68	11	96

The Istanbul Metro was constructed in 1989. Currently it has 107 stations, and 104 more stations are under construction.

The Eiffel Tower. The magic total is 99 + 66 + 88 + 11 = 264. Keeping given numbers in bold and adding a row above and a column to the right to present numbers needed to get the sum of 264 in each column and row, let us write numbers suitable for each empty entry:

198	165	253	176	
1, 8	**99**	6, 8	1, 6	165=18=86+61= 81+68+16
66	1,8	8, 9	1, 9	198
1, 9	1, 6	6, 9	**88**	176
8, 9	6, 8	**11**	6, 9	253

Then only two options are possible

18	**99**	86	61
66	81	98	19
91	16	69	**88**
89	68	**11**	96

Entries from Row 1 and Column 1 and from Row 4 and Column 1 written side by side give 1889.

Entries from Row 2 and Column 4 and Row 4 and Column 2 written side by side are 1968.

81	**99**	68	16
66	18	89	91
19	61	96	**88**
98	86	**11**	69

Entries from Row 1 and Column 1 and Row 4 and Column 1 written side by side are 8198.

Entries from Row 2 and Column 4 and Row 4 and Column 2 written side by side are 9186.

From these candidates for the years, only the first two are realistic.

The Eiffel Tower was opened in 1889 and chose its current color in1968.

Countries Rich with Lakes. There are multiple strategies to fill in the Inky. For instance, Cage 5/ can be presented with 5 and 1, then only 2 and 4 are left for Cage 6+. Hence, the only digit 3 is left for the upper Cage 5+, and then 3+2 completes Cage 5+. The process is continued to fill in the entire Inky.

6+ 2	4	5/ 5	1	5+ 3
3/ 3	1	8+ 4	3- 5	2
12x 4	3	1	2	20x 5
6+ 5	3- 2	3	4	1
1	5	24x 2	3	4

There are an estimated 31,752 lakes in Canada larger than 3 km^2.

Sources

www.weforum.org/agenda/2017/08/these-countries-have-the-most-official-languages/

en.wikipedia.org/wiki/Languages_of_Zimbabwe

en.wikipedia.org/wiki/Languages_of_Singapore

en.wikipedia.org/wiki/Palauan_language

lisbdnet.com/what-is-the-national-language-of-australia

www.weforum.org/agenda/2017/08/these-countries-have-the-most-official-languages

www.nationalgeographic.org; acutrans.com/why-doesnt-the-united-states-have-an-official-language/

www.worldatlas.com/articles/top-10-interesting-facts-about-palau

www.britannica.com,

en.wikipedia.org/wiki/Australia

earthhow.com/7-major-tectonic-plates/

www.usgs.gov/faqs/what-earthquake-and-what-causes-them

www.britannica.com/question/Why-are-there-tectonic-plates
www.businessinsider.com/most-earthquake-prone-countries-in-the-world-2018
diabetes.org/blog/history-wonderful-thing-we-call-insulin
penntoday.upenn.edu/news/100-years-insulin, www.nobelprize.org/prizes/medicine
www.legalzoom.com/articles/top-5-accidental-inventions-discoveries
www.travelandleisure.com/attractions/landmarks-monuments/eiffel-tower-facts
www.history.com/news/10-things-you-may-not-know-about-the-eiffel-tower
www.onthegotours.com/blog/2021/12/interesting-facts-about-istanbul/
www.markfarrar.co.uk/othmsq01.htm
lisbdnet.com/how-many-lakes-are-in-canada
sudoku.com/how-to-play/the-history-of-sudoku/ LOGIC PUZZLES

5.4 Logic Puzzles

Countries of the Past

The powerful Austro-Hungarian Empire combined Austria, Hungary, and some parts of Czechia, Poland, Italy, Romania, and the Balkans. Fifteen new countries were formed after collapsing the most powerful communist nation in the world, Union of Soviet Socialist Republics or USSR. Syria and Egypt joined together to form the United Arab Republic or UAR and abandoned the alliance later.

Problem. Austro-Hungarian Empire, the UAR, and the USSR with their capitals in Budapest, Cairo, Moscow, and Vienna, were formed in 1867, 1922, and 1958 and dissolved in 1918, 1961, and 1991. Austria-Hungary with its two capitals in Vienna and Budapest was the first one to be established, while the UAR with its capital that is not Moscow existed the least number of years compared to the other countries considered there. How long did the USSR last? What was its capital?

All names are given in alphabetical order and years are in increasing order.

Answer. 69

Capitals of Hawaii

A total of 137 volcanic islands of the Hawaiian archipelago span for over 1500 miles in Oceania about 2000 miles from the US mainland in the Pacific Ocean. Hawaii is the latest state to join the Union becoming the only US state outside North America, the only archipelago state, and the only tropical state. Prior to becoming a territory of the United States in 1898 and the State of Hawaii in 1959, Hawaii was the Kingdom of Hawaii until January 17, 1893, and then the Republic of Hawaii.

Problem. Five sites, Hilo, Honolulu, Kailua-Kona, Lahaina, and Waikiki, served as the capital of Hawaii from 1795–1796, 1796–1803, 1803–1812, 1812–1820, 1820–1845, and from 1845 until now. What is its current capital? List the time periods during which each site served as the capital of Hawaii if:

1. Kailua-Kona was the capital of the Kingdom of Hawaii one year longer than Hilo, which was the capital of the Kingdom of Hawaii six years longer than Waikiki.
2. Honolulu served as the capital twice, as the capital of the Kingdom of Hawaii one year longer than Kailua-Kona and then after Lahaina.

Time period	Capital
1795–1796	
1796–1803	
1803–1812	
1812–1820	
1820–1845	
since 1845	

All names are given in alphabetical order and years are in increasing order.

Answer. Honolulu

US Statehood

Thirteen British colonies signed the Declaration of Independence from the Kingdom of Britain giving birth to a new country, the United States of America, on July 4, 1776. Other territories and states have joined the Union expanding it from the original 13 to 50 states, a federal district, and 14 territories. Each US state is distinguished not only by its size, climate, and heritage but also by its state flag, anthem, motto, flower, bird, and animal.

Problem. Alabama, Indiana, Kansas, Mississippi, and New Mexico joined the Union in 1816, 1817, 1819, 1861, and 1912. They chose camellia, magnolia, peony, sunflower, and yucca as their state flower and cardinal, mockingbird, roadrunner, western meadowlark, and yellowhammer as their state bird. States, birds, and flowers are listed in alphabetical order and years are in increasing order. Complete the following table using the information below. The auxiliary table in Figure 5.19 helps to find the solution.

State	Statehood year	State flower	State bird
Alabama			
Indiana			
Kansas			
Mississippi			
New Mexico			

1. Mississippi, the two states that joined the Union in 1819 and 1861, the state with a peony as its state flower, and the state with a roadrunner as its state bird are the five states.
2. A state that chose a camellia as the state flower but not a western meadowlark as the state bird became a state later than Mississippi. A state with a roadrunner, which is not Indiana, does not have a sunflower as the state flower.
3. A state that achieved statehood in 1816 has a cardinal as the state bird and the one in 1817 chose a magnolia to be its state flower.
4. Kansas chose a sunflower but did not choose a mockingbird, which was chosen by a state that chose a magnolia to be its state flower.
5. A state with sunflower as its state flower became the state later than a state with Alabama's state flower camellia but before the state with yucca as its state flower.

Statehood year	State	State flower	State bird
1816			
1817			
1819			
1861			
1912			

	State					State flower					State bird				
	Alabama	Indiana	Kansas	Mississippi	New Mexico	Camellia	Magnolia	Peony	Sunflower	Yucca	Cardinal	Mocking bird	Roadrunner	Western Meadowlark	Yellow hammer
1816															
1817															
1819															
1861															
1912															
Cardinal															
Mockingbird															
Roadrunner															
Western meadowlark															
Yellowhammer															
Camellia															
Magnolia															
Peony															
Sunflower															
Yucca															

FIGURE 5.19

The Northwest Territory

The Northwest Territory, also called the Old Northwest and the Territory Northwest of the River Ohio, was formed from unorganized western territory of the United States after the American Revolutionary War. The region was ceded to the United States in the Treaty of Paris of 1783. Throughout the Revolutionary War, the region was part of the British Province of Quebec.

Problem. Illinois, Indiana, Michigan, Minnesota, Ohio, and Wisconsin used to be parts of the Northwest Territory before they became states in 1803, 1816, 1818, 1837, 1848, and 1858. Their capitals are Columbus, Indianapolis, Lansing, Madison, Saint Paul, and Springfield. 5.4, 5.7, 6.5, 9.9, 11.5, 12.9 million people live in these states with the areas of 36.4, 44.8, 57.9, 65.6, 86.9, 96.7 thousand square miles. Identify the date of statehood, capital, the area, and the population size of each state if

1. Statehood from the earliest year: the state with Columbus as its capital, the state with the smallest area, the state with the largest population size, the state with the largest area, Wisconsin, the state with the smallest population density.
2. The area of Wisconsin is 31.1 thousand sq. mi less than the area of the state with the capital of Lansing while the area of Illinois 13.1 thousand sq. mi more than the area of the state that joined the union 55 years earlier than Minnesota.

3. The state with the largest population density of 257.5 p /sq. mi has the capital of Columbus, while the state with the density of 86.9 p /sq. mi has the capital of Madison. The population density of Minnesota is 62.1 p /sq. mi.
4. Both the area and population size are above their corresponding averages for just one state that joined the Union in 1837, while they are below in Indiana that has Indianapolis as its capital.
5. 1.4 million less people live in Ohio than in the state with its capital Springfield, which is not Michigan.

Remark. The auxiliary table in Figure 5.20 is recommended to find the solution.

Date of statehood	State	Total area	Population	Capital
1803				
1816				
1818				
1837				
1848				
1858				

An Insular Area

An insular area of the United States is a territory under US-associated jurisdiction though it is neither a part of one of its 50 states nor a part of the District of Columbia. People that live there do not have the right to vote in US elections. The word *insular* means *island* in Latin.

Problem. American Samoa, Guam, Northern Mariana Islands, Puerto Rico, and US Virginia Islands are five inhabited insular areas of the United States. Their villages or cities, Charlotte Amalie, Hagatna, Pago Pago, Saipan, San Juan, became their official capitals in 1898, 1899, 1917, and 1947.

Countries and cities are listed in alphabetical order, years are given in ascending order. Using the hints below, list the five US insular areas with their capital and the year when the city became its capital. Use the auxiliary table in Figure 5.21 to find the answers.

1. The cities became the capital in the order from the earliest year: both the capital of Guam and San Juan, the capital of American Samoa, Charlotte Amalie, and the capital of Northern Mariana Islands.
2. Pago Pago was not a capital in 1898, while Saipan became the capital 30 years after US Virginia Islands chose its current capital.

	State						Population						Total area sq. mi						Capital					
	Illinois	Indiana	Michigan	Minnesota	Ohio	Wisconsin	5.4	5.7	6.5	9.9	11.5	12.9	36.4	44.8	57.9	65.6	86.9	96.7	Columbus	Indianapolis	Lasing	Madison	Saint Paul	Springfield
1803																								
1816																								
1818																								
1837																								
1948																								
1858																								
Columbus																								
Indianapolis																								
Lansing																								
Madison																								
Saint Paul																								
Springfield																								
36.4																								
44.8																								
57.9																								
65.6																								
86.9																								
96.7																								
5.4																								
5.7																								
6.5																								
9.9																								
11.5																								
12.9																								

FIGURE 5.20

	American Samoa	Guam	Northern Mariana Islands	Puerto Rico	U.S. Virginia Islands	Charlotte Amalie	Hagatna	Pago Pago	Saipan	San Juan
1898										
1898										
1899										
1917										
1947										
Char. Amalie										
Hagatna										
Pago Pago										
Saipan										
San Juan										

FIGURE 5.21

The Tallest Towers

CN Tower, Ostankino Tower, Sky Tower, Stratosphere Tower, and Tokyo Skytree located in Canada, Japan, New Zeeland, Russia, and the United States of America are among the tallest towers in the world, in the Western or Southern Hemispheres, in Europe, or the United States. All sets of items are listed in alphabetical order. The towers were built in 1967, 1976, 1996, 1997, 2012 and, being 1076, 1149, 1772, 1815, 2080 feet tall, are ranked as 1, 3, 4, 17, and 25 in the list of the tallest towers.

All number sets are listed in increasing order. Using the hints below, describe all mentioned towers. The auxiliary table in Figure 5.22 is recommended.

1. The years when the tallest tower in the world and the tallest observation tower in the United States were built as well as the heights of the tallest structure in Europe and in the Southern Hemisphere have the same number of integer divisors.
2. Either the year when the tallest freestanding structure in New Zealand, which is in the Southern Hemisphere, was built or its height is a prime number.

		Tower					Year					Country					Recognition				
Rank	Height	CN Tower	Ostankino	Sky Tower	Stratosphere	TokyoSkytree	1967	1976	1996	1997	2012	Canada	Japan	New Zealand	Russia	USA	Europe	Southern	USA	Western	World
1	2080																				
3	1815																				
4	1772																				
17	1149																				
25	1076																				
Recognition	Europe																				
	Southern																				
	US																				
	Western																				
	World																				
Country	Canada																				
	Japan																				
	NZ																				
	Russia																				
	USA																				
Year	1967																				
	1976																				
	1996																				
	1997																				
	2012																				

FIGURE 5.22

3. The year when CN Tower was built has the largest number of divisors among all years, while the height of Tokyo Skytree in Japan has the largest number of divisors among all numbers, which is twice the number of divisors of the height of Canadian CN Tower and six times the number of divisors of the year when Ostankino Tower was built and the number that represent the height of the tallest observation tower in the United States built after CN Tower.
4. Tokyo Skytree, Sky Tower, and Ostankino Tower are among the shortest, the highest, the oldest, and the youngest, or both.
5. The tallest European tower located in Russia was built before all other mentioned towers, though it is not the shortest one.

Rank	Tower	Height	Year of construction	Recognition	Country
1					
3					
4					
17					
25					

Suspension Bridges

The deck of a suspension bridge is hung below two or more suspension cables passing over vertical towers and securely anchoring at the ends. Simple suspension bridges have been found to be built in the 15th century. One of the first modern suspension bridges, Menai Suspension Bridge over the Menai Straits connecting the island of Anglesey with the mainland of Wales was opened in 1826.

There are a lot of different types of suspension bridges and most of them are unique and notable. The 1915 Çanakkale suspension bridge that spans the Dardanelles connecting two continents is the world's longest suspension bridge. Previously the world's longest bridge, Union Chain Bridge or Union Bridge that goes over the River Tweed is the world's oldest suspension bridge still carrying road traffic. Another bridge that used to be the world's longest is Golden Gate Bridge, which also used to be the world's tallest bridge. The bridge has retained this title but only in its country. Having a width of 67.3 meters, Rod El Farag Bridge holds the Guinness World Record for the world's widest suspension bridge. Zhangjiajie Glass Bridge is the world's longest and highest glass-bottomed bridge connecting two mountains and offering an incredible view for brave visitors.

Problem. The five bridges, Golden Gate, the 1915 Çanakkale Bridge, the Rod El Farag Bridge, the Union Chain Bridge, and Zhangjiajie Glass footpath, are

	Bridge					Location					Main span in meters				
	1915 Canakkale	Golden Gate	Zhangjiajie Glass	Rod El Farag	Union Chain	China	Egypt	Great Britain	Turkey	USA	137m	430m	540m	1280m	2023m
1820															
1937															
2016															
2020															
2022															
137 m															
430 m															
540 m															
1280 m															
2023 m															
China															
Egypt															
GB															
Turkey															
USA															

FIGURE 5.23

located in China, Egypt, Great Britain, Turkey, or the USA. They were opened in 1820, 1937, 2019, 2020, or 2022. Their longest spans are 137 m, 526 m, 540 m, 1280 m, and 2023 m. Find the location of each bridge, its length of main span, and the date of its completion. Fill in the table in Figure 5.23.

All names are given in alphabetical order and numbers in increasing order.

1. The average length of the main spans of two bridges constructed in the 21st century is 63 m less than four times the span length of the Union Chain Bridge, though four times this average is 83 m less than the span length of the third bridge completed in the 21st century, which is not in Egypt. None of the mentioned here bridges was built in the USA.
2. The oldest, the longest, the Golden Gate Bridge, the Chinese Zhangjiajie Glass bridge, and the bridge with the median span of all five bridges describe the five bridges.
3. The difference between the span lengths of the newest and the oldest bridges is 1886 m. One of them is located in Great Britain, the other bridge is the 1915 Çanakkale Bridge.
4. The bridge that has the median span was not built in the year which is the median year of all given years.
5. The Golden Gate bridge was built neither in Turkey nor in the 21st century.

Date of completion	Bridge	Location	Length of main span
1820			
1937			
2019			
2020			
2022			

Countries Rich with Islands

Can you guess which countries have the most islands? Yes, three Scandinavian or Nordic countries, Sweden, Norway, and Finland have the most islands in the world. Sweden's thousands of islands are located along its eastern coastline from the far north to the deep south. About 1000 of them are inhabited and 24,000 islands are open to the public and can be visited without permission. The Gothenburg Archipelago made up of over 8000 islands and islets has been ranked as the seventh most beautiful natural wilderness area in the world by CNN Travel.

Problem. Find the number of islands and islets in Sweden if this number is presented by all digits from the last row of the number pyramid puzzle written side by side and shown in Figure 5.24.

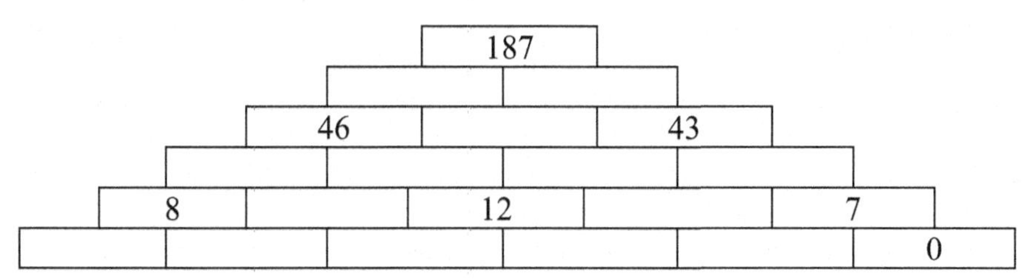

FIGURE 5.24

Can you solve the number pyramid puzzle without having a digit in the very last row?

Remark. Each number in a *number pyramid* is the sum of two numbers directly beneath it. All numbers are nonnegative.

Answer: 267,570

Countries of the World

The number of countries has changed over time and has increased from 77 countries in 1900. The current list of countries considers recognized sovereign states, associate states, and entries which claim sovereignty. Each country has its own culture, history, and heritage.

Problem. Find the current number of countries in the world, if this number is decoded in the four-digit crack-the-code puzzle:

1234 one digit is correct but wrong placed.

8765 one digit is correct and well placed.

8719 two digits are correct but wrong placed.

9061 three digits are correct but wrong placed.

Digits in the number cannot be repeated.

Answer: 195

The Oldest Border

Borders between countries have changed over time. The sixth-smallest European country located in the Pyrenees Mountains, Andorra, has the oldest 120 km border with France and Spain. Charlemagne took the region from the Muslims in 803 granting the country independence. Andorra has had strong ties with Catalonia, Spain, and accepted Catalan as its official language. Most Andorrans live in its capital, Andorra la Vella, translated from Catalan as *Andorra the Old*.

The Madriu-Perafita-Claror valley in Andorra was designated as a UNESCO World Heritage Site in 2004 for its astonishing glacial landscapes and unforgettable views.

Problem. Find the year when the world's oldest Andorra's border with France and Spain fixed in a feudal charter signed on September 8, if the year is decoded in the four-digit crack-the-code puzzle:

2734 two digits are correct but incorrectly placed.

3508 one digit is correct and well placed.

1963 one digit is correct and well placed.

2567 two digits are correct but incorrectly placed.

The digit 0 is placed as the first digit of a year if the year was before the 11th century. Digits in the number cannot be repeated.

Answer: 1278

Solutions

Countries of the Past. It is easy to see that Moscow was the capital of USSR and that the Austro-Hungarian Empire existed from 1867 until 1918 because the other two numbers 1922 and 1958 are greater than 1918. Checking the differences 1961 – 1922 = 39, 1961 – 1958 = 3, 1991 – 1922 = 69, and 1991 – 1958 = 33, leads to 3 as the lowest number.

The Austro-Hungarian Empire established in 1867 and in 1918, the United Arab Republic existed from 1958 to 1961, though Egypt kept the name United Arab Republic for another decade. USSR lasted from 1922 until 1991.

Capitals of Hawaii. Rows 1, 2, and 4 are filled based on information given in Hint 1. The rest follows from Hint 2.

Time period	Capital
1795–1796	Waikiki
1796–1803	Hilo
1803–1812	Honolulu
1812–1820	Kailua-Kona
1820–1845	Lahaina
since 1845	Honolulu

Honolulu was the capital of the Kingdom of Hawaii from 1803 to 1812 and has served as the seat of the Provisional Government of Hawaii, the capital of the Republic of Hawaii, the Territory of Hawaii, and the state of Hawaii since 1845.

US Statehood. The table is completed following the hints of the problem. For instance, Hint 1 tells that Mississippi did not join the Union in 1819 and 1861 (that leaves 1816, 1817, and 1912 as possible years), does not have a peony as its state flower, and does not have a roadrunner as its state bird. States that have a peony as the state flower or a roadrunner as the state bird did not join the Union in 1819 and 1861 either. The state that has a peony as the state flower does not have a roadrunner as the state bird. These options should be crossed out from the corresponding cells in the table. Other hints help to complete the table.

State	Statehood year	State flower	State bird
Alabama	1819	Camellia	Yellowhammer
Indiana	1816	Peony	Cardinal
Kansas	1861	Sunflower	Western Meadowlark
Mississippi	1817	Magnolia	Mockingbird
New Mexico	1912	Yucca	Roadrunner

The Northwest Territory. The problem can be solved using the following steps:

- Calculate the density of a state as the ratio between the area and the population, and present the density in the last column added to the table.
- Calculate the average density, the average area of the states, and the average population of the states, and present them in the last row added to the table.
- Compare the total area of the states with the average area, and add the result next to the area of the state.
- Compare the total population of each state with the average population and add the result next to the total population of the state.
- Follow the problem hints to complete the table.

State	Date of statehood	Total area	Population	Capital	Density p/mi
Illinois	1818	57.9 below average	12.9 above	Springfield	222.8
Indiana	1816	36.4 below	6.5 below	Indianapolis	179.5
Michigan	1837	96.7 above	9.9 above	Lansing	102.2
Minnesota	1858	86.9 above	5.4 below	Saint Paul	62.1
Ohio	1803	44.8 below	11.5 above	Columbus	257.5
Wisconsin	1848	65.6 above	5.7 below	Madison	86.9
Average		64.7	8.65		

An Insular Area. The second hint tells us that Saipan became the capital in 1947, and US Virginia Islands had its current capital in 1917. Then Pago Pago became the capital in 1899. The first hint gives the remaining information needed.

American Samoa	Pago Pago	1899
Guam	Hagatna	1898
Northern Mariana Islands	Saipan	1947
Puerto Rico	San Juan	1898
US Virginia Islands	Charlotte Amalie	1917

The Tallest Towers. The relation between the rank and height of a tower is obvious. Therefore, they take the first two columns and are considered together. Let us find how many divisors each number has using the statement, that

- if the prime factorization of a number A is $A = a^l \cdot b^m \cdot c^n$, then the number of its divisors is $(l + 1) \cdot (m + 1) \cdot (n + 1)$

and set up a table based on this information.

Height	#of divisors	Year of construction	#of divisors
$1076 = 2^2 \cdot 269$	6	$1967 = 7 \cdot 281$	4
$1149 = 3 \cdot 383$	4	$1976 = 2^3 \cdot 13$	16
$1772 = 2^2 \cdot 443$	6	$1996 = 2^2 \cdot 449$	6
$1815 = 3 \cdot 5 \cdot 11^2$	12	1997	prime
$2080 = 2^5 \cdot 5 \cdot 13$	24	$2012 = 2^2 \cdot 503$	6

Hint 1 states that the year when the tallest tower in the world was built is either in 1996 or 2080 and the year when the tallest observation tower in the United States was built is either in 1996 or 2080. The height of the tallest structure in Europe is either 1076 ft or 1772 ft, and the height of the tallest structure in the Southern Hemisphere is either 1076 ft or 1772 ft.

Other problem hints and the table that contains the number of divisors lead to the conclusion presented in the table.

Rank	Tower	Height in ft	Year of construction	Highlights	Country
1	Tokyo Skytree	2080	2012	The tallest structure in the world	Japan
3	CN Tower	1815	1976	The tallest structure in western hemisphere	Canada
4	Ostankino Tower	1772	1967	The tallest structure in Europe	Russia
17	Stratosphere Tower	1149	1996	The tallest observation tower in the United States	United States of America
25	Sky Tower	1076	1997	The tallest freestanding structure in the southern hemisphere	New Zealand

Suspension bridges. Hints and required calculations lead us to the results summarized in the table.

Bridge	Location	Length of main span	Date of completion
The 1915 Çanakkale Bridge	Turkey	2023 m	2022
The Golden Gate	USA	1998 m	1937
The Rod El Farag Bridge	Egypt	540 m	2019
The Union Chain Bridge	Great Britain	137 m	1820
Zhangjiajie Glass footpath	China	430 m	2016

Countries Rich with Islands. To complete the puzzle, consider a small pyramid, e.g.,

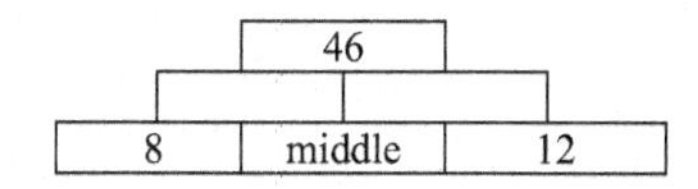

and try to predict the middle number knowing three corner numbers.

It is (46 – (8 + 12))/2 = 13. Fill in all middle numbers using this hint and consider other small pyramids until the given pyramids is filled.

					187					
				95		92				
			46		49		43			
		21		25		24		19		
	8		13		12		12		7	
2		6		7		5		7		0

Sweden has 267,570 islands and islets.

The puzzle can be solved without having 0 in the last row, but there will be several possible answers, e.g., 267,570, 176,661, 085,752. The answer is not unique in this case.

Countries of the World. The first two numbers, which contain eight digits 1234 and 8765 that satisfy two hints, leads to the conclusion that digits 0 and 9 are included in the number, because a four-digit number is being hidden. The third number 8765 has two digits that are incorrectly placed. Hence neither 8 nor 7 are in the secret number because if at least one on them were in the number, then there would have been at least one well-placed answer. Thus, 1 and 9 are in the number to be guessed. Moreover, comparison of the second and fourth hints tells us that 6 is not in the secret number. Combining our outcomes, we can say that:

- 0 can be at the first, third, or fourth place.
- 1 cannot be at the first place.
- 9 can be on the second or third place.
- 5 takes the fourth place.

Putting everything together, the secret number is 0195.

There are 195 countries in the world. 193 of them are UN members and 2 countries, Holly See (Vatican) and State of Palestine are observer states.

The Oldest Border. The lock puzzle can be solved in different ways. See one of them. The first three numbers contain all ten digits with 3 repeated in all three numbers. Then there are two options:

1. Digit 3 is among the digits of the year. Hence, one digit is among 2, 7, or 4, and the next two numbers cannot give any digits because their digit 3 has been already counted on. This leads to a two-digit number and a subsequent contradiction. Thus, this option does not work.
2. Digit 3 is not among the year digits. Then the other two digits are among 2, 7, and 4, one digit is 5, 0, or 8, and one digit is 1, 9, or 6. Comparing

second and third numbers with the fourth number calls for excluding 5 and 6 from the number of the year, because their positions remain the same, but the statement changes from *well placed* to *wrong placed*. That leaves us with six (instead of four needed) potential digits for the year:

- either 1 at the first place or 9 at the second place,
- either 0 at the third place or 8 at the fourth place,
- 2 at the second or third place,
- 7 at the first, third, or fourth place.

Considering these options, we can obtain two possible years 1278, 7928, or 7902. The last two years are from the future.

The world's oldest border of Andorra with France and Spain has remained the same since September 8, 1278.

Sources

lithub.com/the-oldest-the-longest-the-weirdest-a-brief-history-of-land-borders/

www.britannica.com/place/Andorra

www.worldatlas.com/articles/which-countries-have-the-most-islands.html

www.statista.com/chart/15364/the-estimated-number-of-islands-by-country

www.britannica.com/technology/suspension-bridge

en.wikipedia.org/wiki/Suspension_bridge

en.wikipedia.org/wiki/List_of_tallest_towers

en.wikipedia.org/wiki/Insular_area

en.wikipedia.org/wiki/Hawaii

Index

Printed in the United States
by Baker & Taylor Publisher Services